Deductive Logic

*An Introduction to Evaluation Techniques
and Logical Theory*

By D. S. CLARKE, JR.

Southern Illinois University Press
Carbondale and Edwardsville
Feffer & Simons, Inc.
London and Amsterdam

To My Mother and Father

Library of Congress Cataloging in Publication Data

Clarke, David S.   1936-
    Deductive logic.

    Bibliography: p.
    1. Logic. 2. Deontic logic. 3. Modality (Logic)
I. Title.
BC71.C53   162   73-10459
ISBN 0-8093-0657-3

Copyright © 1973 by Southern Illinois University Press
All rights reserved
Printed by offset lithography in the United States of America

CONTENTS

	Page
Preface	vii
I. Introduction	1
1. The Variety of Inferences	1
2. Validity and Soundness	4
3. Sentences and Propositions	5
4. Object Language and Meta-Language	8
II. Sentence Logic	13
5. Logical Form	13
6. Negation, Conjunction, and Disjunction	15
7. Implication, Equivalence, and the Stroke Function	21
8. Evaluation of Inferences	27
9. General Decision Procedure	33
10. Disjunctive Normal Form	37
III. The Sentence Calculus	44
11. Rules of Inference	44
12. Primitive Rules Governing Implication	46
13. Rules for Negation, Conjunction, and Disjunction	53
14. Derived Rules of Inference and Applications	58
15. Categorical Proofs	63
IV. Predicate Logic: Basic Concepts	68
16. Terms	68
17. Logical Representation of Analyzed Sentences	73
18. Complex Quantificational Schemata	78
19. Representation of Relational Sentences	83
V. Decision Procedure for Predicate Logic	91
20. Decision Procedure for Finite Domains	91
21. Aristotelian Logic	97
22. Distributive Normal Forms	104
23. Decision Procedure for Predicate Logic	110
24. Relational Predicates	116
25. The Principle of Bivalence	122
VI. The Predicate Calculus	130
26. Rules for the Universal Quantifier	130
27. Rules for the Existential Quantifier	135
28. Applications of the Predicate Calculus	139
29. Predicate Variables and Identity	146
30. Syntax and Semantics	153

VII.	Modal Logic	161
	31. The Modal Trichotomies	161
	32. Interpretations of the Alethic Modalities	165
	33. Semantic Decision Procedure	171
	34. The Alethic Modal Calculus	178
VIII.	Imperatives and the Deontic Modalities	190
	35. Imperative Inferences	190
	36. Mixed Inferences	197
	37. The Calculus Applied to Imperatives	204
	38. Deontic Sentences	211
	39. Decision Procedure for Deontic Logic	217
	40. The Deontic Calculus	221
Appendix. Summary of Syntactic Principles and Rules		232
Bibliographical References		237
Index		242

PREFACE

As the branch of mathematics called "mathematical logic" logic has been developed as a means by which the mathematician can study the properties of systems of axioms. The methods used here are necessarily rigorous and to understand them requires some degree of mathematical sophistication. Logic is also a means by which all of us can evaluate as good or bad the reasoning of ourselves and others, no matter what the subject matter. As such it is of interest to the general student lacking a developed background in mathematics and perhaps a special aptitude for using its methods. It is this latter audience to which this text is directed. Its exposition is therefore informal, with emphasis on explaining the basic concepts and procedures of modern symbolic logic rather than on developing a rigorous formal system and providing proofs of its properties. The presentation is exact enough, however, to serve as a reliable foundation for advanced work by those with more specialized interests.

Courses covering this type of material are almost invariably offered within philosophy departments. Here there are usually two basic courses offered. First comes a first-level (freshman or sophomore) course treating logic as a practical tool for reasoning more effectively and usually including inductive logic. Later, usually at the junior level, is a course devoted exclusively to the techniques of modern symbolic logic as applied to deductive reasoning. This text is intended to meet the needs of this second course, though students with some earlier preparation in logic will find much in the way of review in the early chapters. The fact that it is a course offered to philosophy students and serves to introduce them to logic as the "language of philosophy" has strongly influenced the selection of topics. It has been the reason for raising at various points in the text some of the central philosophical problems arising from the applications and assumptions of modern logic. Notes are provided at the end of each chapter referring to the more important (and accessible) sources in which these problems are discussed. These notes also serve to acquaint the reader with alternative formal methods and supplementary material in the number of fine texts now available.

A major departure from almost all available logic texts is the extension in the final chapter of logical techniques of evaluation to reasoning about our conduct. That a form of reasoning central to everyday life and the formulation and application of the law has been largely ignored thus far is nothing short of a scandal, and it is high time we begin to set matters right. This application of logic has been marked by controversy, to be sure, and there are still issues on which there is disagreement. Nevertheless, sufficient progress has been made in this field to justify a text-book exposition, though the reader should be warned that what is stated here is of a more provisional nature than in the earlier chapters on relatively established topics.

The text can be adapted for a variety of different types of courses. As is indicated in the text, Sections 10 and 22-24 may be omitted for students with no previous background in symbolic logic. Sections 25, 30, and 32 concentrate on philosophic topics, and may be omitted by those not wanting to stray from the central problems of evaluation. The first six chapters could constitute a one-semester course introducing symbolic logic, while the last two chapters supplemented by selected readings on logical theory of the kind referred to in the notes could constitute a second course aimed towards students with strong philosophic interests.

Exercises are provided at the end of sections in which formal procedures are developed. These are designed to review the chief points made in the text and not to introduce new material.

This text has been used in various forms in my logic classes at Southern Illinois University over the past seven years. Its development owes greatly to the criticisms and suggestions of colleagues and students at the University, especially those of James Magruder, Michael Audi, Richard Behling, Charles Richardson, and Jonathan Seldin. I am also grateful to Gary Iseminger of Carleton College for reading through an earlier version and making a number of suggestions that lead to substantial alterations. Finally, I would like to thank Richard Delaney of Research and Projects at Southern Illinois for his encouragement and assistance, Sharon Patterson and Joanne Hinkle for the typing of the manuscript, and Walter Kent and Beatrice Moore of the University Press for their help with its final preparation.

D. S. Clarke, Jr.

Carbondale, Illinois
June 12, 1973

I. INTRODUCTION

1. The Variety of Inferences

Very rarely in our every-day and public lives do we communicate to others by using isolated sentences. In conversation, letters, scientific papers, official directives, novels, etc., sentences are almost invariably combined to form what is called a discourse. Often the only relation between what the various sentences express is derived from the common topic of the discourse. The conversation may be about the weather, and this topic will relate the different sentences used in the conversation; the scientific paper may be about the structure of a certain enzyme, a topic that gives unity to the paper and its constituent sentences. But sometimes the relation between sentences can be stronger than this, a relation that is signalled by the occurrence of such words as 'therefore', 'hence', 'since', and 'because'. Here one sentence is said to "follow from" or be "justified by" others. When this occurs we have what is called an inference. The sentence being derived or justified we call the conclusion of the inference, those from which it is derived the premisses. An inference may be defined, then, as a type of discourse in which the constituent sentences are related in such a way that one of them, the conclusion, is said to be justified by others, the premisses.

Logic we shall understand as the science that determines whether or not the conclusion of an inference ought to be inferred from the premisses. It is thus being understood as what the American philosopher-logician Charles Peirce called a "normative science," a science that evaluates what ought to be done in contrast to a natural science describing what is the case. To be sure, the term 'logic' has often been used by philosophers in a sense much broader than this. There is the "dialectical logic" of Hegel and Bradley and the "logic of ordinary language" of Wittgenstein and his followers. We shall ignore for the most part, however, these generalized conceptions of the subject in order to concentrate on that central part boasting a continuous history from Aristotle to the present.

There are a variety of ways that the premisses of an inference can be related to its conclusion, and these differences mark off different types of inferences. Among the more important are practical, inductive, and deductive inferences. For each of them there is a separate field of logic in which they are evaluated.[1]

Practical Inferences. These are inferences in which a course of action is inferred from one premiss about a desired end and another stating what is necessary to attain this end.[2] An example of such an inference would be

> Jones should go to college, since he wants to become a junior executive, and in order to become one he must go to college.

Here the conclusion, 'Jones should go to college', is said to be justified by the premisses 'Jones wants to become a junior executive' and 'In order to become a junior executive Jones must go to college'. This is a type of inference commonly used in recommending a course of action either to oneself or others on the basis of a desired end and knowledge of means necessary to attain that end. This is a "weak" type of inference in the sense that the truth of the premisses is not sufficient to guarantee the conclusion that an action should be pursued. There may be other consequences of necessary means to an end which counter-balance the desirability of the end. Going to college, for example, may be unpleasant or costly enough for Jones to outweigh for him the benefits of becoming a junior executive. The conclusion would not, therefore, hold.

Inductive Inferences. Another important type of inference is that in which the premisses are singular sentences reporting the outcomes of direct observation and the conclusion an empirical generalization. An example of an inductive inference would be

> This piece of copper conducts electricity. This other piece of copper also conducts electricity, and so do all the others that have been observed. Hence, all copper must conduct electricity.

In this inference the conclusion, 'All copper conducts electricity', is said to follow from the separate premisses reporting that observed pieces of copper have conducted electricity. As before for practical inferences, the relation between sentences is such that the truth of the premisses does not insure the truth of the conclusion. Instead, the premisses serve to strengthen our confidence in the conclusion, the degree of this confidence depending on the manner in which the reported observations have been made.[3]

Deductive Inferences. These are inferences with a "strong" relation between premisses and conclusion where the claim is made that for the premisses to hold guarantees that the conclusion does also. Included among deductive inferences would be those like the following:

>   This piece of metal conducts electricity, because
>   all copper conducts electricity and this metal is
>   copper.

in which the truth of the premisses 'All copper conducts electricity' and 'This metal is copper' does insure the truth of the conclusion 'This piece of metal conducts electricity'. Also included, as we shall see in Chapter VIII, are inferences whose constituent sentences are in the imperative mood, those such as

>   Either mail the letter or burn it. But don't
>   burn it. Therefore, mail the letter.

in which to obey the premisses insures that the conclusion, 'Mail the letter', is obeyed.

There are, of course, other types of inferences besides the three just cited. The lawyer argues for the innocence of his client on the basis of the existing evidence and the laws of the community. The legislator argues for the need of a new law because of some inequity in society that cannot be corrected by the present laws. Neither of the inferences they formulate would be properly classified under the types we have listed, and there are undoubtedly many more. But within this great variety of inferences a general distinction can be made between deductive inferences and the others. For a deductive alone is there the strong relation between premisses and conclusion on the basis of which the claim can be made that the holding of the premisses guarantees the conclusion holding also. For both the practical and inductive inferences the relation is weaker, and the claim cannot be made. In general, all inferences other than the deductive are capable of being defeated by a single case or further consequence that may have been overlooked or that the future may bring which proves that the conclusion does not hold despite the holding of the premisses. Deductive inferences are alone immune from the possibility of such defeat.

Our attention in the following chapters will be restricted exclusively to those inferences for which a claim for the strong relation is made, to deductive inferences. 'Inference' for us in what follows will thus always be understood as 'deductive inference', and 'logic' is to be understood as 'deductive logic', the science in which the strong claim made by deductive inferences is evaluated as being justified or not justified.

All of this suggests that logic is exclusively a study of inferences. This is not true, as we shall see in the last section of the following chapter. Logic as we are understanding the subject is <u>primarily</u> a theory of inference, but not exclusively so. The inference is its starting point, its immediate subject matter. But logic has also the extended aim of distinguishing complex sentences that express logical (necessary) truths from those that do not. We concentrate our discussion in this chapter and the next on inferences because this is the starting point from which the more general theory applicable to sentences is developed.

2. Validity and Soundness

We have just seen that what distinguishes a deductive inference from other types is the implicit claim that if the premisses hold so must the conclusion also. If the claim is justified, if it is impossible for the conclusion to not hold if the premisses do, an inference is to be evaluated as valid; if not, the inference is invalid. In the chapters that follow we shall be occupied with developing procedures by which evaluation of different classes of deductive inferences can be made, primarily those where for a premiss or conclusion to hold is for it to be true.

Such evaluation is conducted independently of the consideration of additional claims that may accompany the formulation of inferences. For besides the general claim that marks a deductive inference, a speaker or writer may further claim that the conclusion is true because of premisses accepted as true, and may use the inference to convince the hearer or reader of this truth. The inference is then used with what we shall call argument force. Such would be an inference used by a lawyer in defending his client before a jury or a debater seeking to refute his opponent. Inferences as used by philosophers are also usually of this kind. The intention of those who argue for God's existence, for example, is to convince the reader of the truth of the conclusion that God exists, not simply to claim the conclusion follows from the premisses. That an inference is intended to have the force of an argument can usually be detected by the context in which it occurs, the tone of voice with which it is uttered (if spoken), or by the order of premisses and conclusion. If the conclusion comes first, we are almost invariably in the presence of an inference with argument force, as in 'That man killed Smith, since it was either he or Jones, and Jones didn't do it', where the conclusion is clearly intended to be asserted as true on the basis of the accepted truth of the premisses.

If an inference is valid and if the premisses are in fact true, then the conclusion is also true. The claim accompanying arguments is then justified and the inference is said to be sound. A sound inference must, therefore, be valid. But not every valid inference is sound. The following is an example of a valid inference:

> Either Lincoln was assassinated or he survived
> his wife. But Lincoln was not assassinated.
> Therefore, he survived his wife.

But since its conclusion is false because of a false second premiss, the inference is unsound.

But to determine whether or not the constituent premisses of an inference are actually true or false is not within the province of the logician. This can only be determined with reference to the type of evidence that the scientist, historian, lawyer, etc., employs to establish their assertions. The logician cannot therefore assess the entire claim to soundness accompanying the use of inferences with argument force. He restricts himself to establishing whether or not <u>if</u> the premisses <u>were</u> true, the conclusion <u>would be</u> also. The act<u>ual</u> truth or <u>falsity</u> of the premisses c<u>an only</u> be assessed by others.

The determination of validity or invalidity can be applied, however, to a second use of inferences, their use with what may be called <u>hypothetical force</u>, a force indicated by phrases such as 'Let us suppose that...' or 'Assume that...' occurring before the premisses. This is the use commonly found in science where a hypothesis whose truth or falsity is in doubt is taken as a premiss from which conclusions are drawn that can be related to direct observation. The observation can then serve as a means of confirming or falsifying the hypothesis. Inferences with hypothetical force are also typical in mathematics where axioms are assumed as premisses in order to trace out theorems as conclusions. That the axiom and theorems are true is never claimed by the mathematician; he claims only that if they were true so would be the conclusion. Now this claim logic is equipped to evaluate. If the inference in question is valid it is justified; otherwise, it is not.

Deductive logic thus has a direct application to an important use of inferences. Its evaluation procedures also apply to those inferences used with argument force to persuade us of a certain conclusion. The claim here will not stand unless the inference used is valid, though this validity does not in itself establish the claim.

3. Sentences and Propositions

The basic unit of human communication is the sentence. Before proceeding to evaluate inferences in the next chapter we consider sentences as isolated from the context of inferences in which they may occur and develop some key distinctions.

We distinguish first of all a <u>sentence</u> as a type of expression from its different occurrences or <u>sentence-tokens</u>. If we are dealing with communication through speech, the sentence-tokens are spoken utterances; if with writing, they are inscriptions, the particular configurations of marks on a page which we interpret in reading. To illustrate the distinction, we list two tokens of the same sentence:

1) It is raining.
2) It is raining.

1) and 2) are different inscriptions; they occupy different portions of this page. But they share a common form and are instances of the same type of expression, that is, they are tokens of the same sentence. Sentence-tokens alone are heard or observed. We never see or hear a sentence, it being instead an abstraction needed to explain our recognition that different tokens such as 1) and 2) are of the same type. It is, however, an indispensable abstraction, and one we shall continue to use. Parallel to the distinction between sentence-tokens and sentences is that between <u>word-tokens</u> and <u>words</u>. There are several tokens of 'the' observable on this page, as contrasted to the one word of which they are occurrences. Again the word is an abstract form or type in terms of which we recognize the particular inscriptions.

Another basic distinction is that between a sentence as a type of linguistic expression and the <u>proposition</u> that it expresses for an interpreter (a hearer or reader) of a token of the sentence. It is clear that what is said by a variety of sentences can be the same for a person. What is expressed by sentences of different languages such as

3) It is raining
4) Il pleut

is the same for one who understands both English and French. This can also occur for different sentences within the same language, e.g., the English sentences

5) John gave the ring to Mary.
6) Mary was given the ring by John.
7) The eldest son of the Smiths gave the ring to Mary. (Assuming John is the Smiths' eldest son.)

The linguistic forms instantiated here are clearly different, and yet they all "say the same thing." The term 'proposition' simply stands for what sentences like 3) and 4) and 5)-7) express in common to one who understands them.[1] Notice that the proposition expressed by a sentence is not equivalent to its meaning. If John is known to be the Smiths' eldest son by the interpreter of 7), then it will express the same proposition for him as 5). But clearly the meanings of 5) and 7) are different due to the difference in meaning between the expressions 'John' and 'the eldest son of the Smiths'. The same proposition can thus be expressed by sentences that differ in meaning.[2]

It is to propositions that the adjectives 'true' and 'false' should be ascribed, since the particular linguistic formulation of a certain utterance or inscription would seem irrelevant to whether we judge it true or false. If we observe it to be raining outside, then an utterance of 'Il pleut' will be judged true as well as an utterance of its corresponding English sentence. And if we know for some reason that it was Alice whom John gave the ring to, not Mary, then utterances of the three sentences 5)-7) will alike be judged false. The conditions under which an utterance of a sentence is judged true or false are thus independent of the particular linguistic form of this sentence. What is alone relevant is what is expressed by the sentence, the proposition. It is, however, often very convenient to speak as if it is sentences to which truth and falsity are ascribed, and we shall often avail ourselves of this expedient in what follows. Thus, '"It is raining" is true' is much more convenient than the cumbersome 'The proposition expressed by the sentence "It is raining" is true'. Yet the latter is the correct form, and the former, when used here, should always be understood as an ellipsis for it.

Sentences are commonly used to express propositions that are asserted or judged true or false relative to some particular time and place, or what we shall call a <u>referent occasion</u>. The referent occasion clearly makes a difference in determining what judgment a hearer makes. An utterance of 'It is raining', for example, said on January 29th in New York may express a proposition judged to be true relative to that time and locale, but said the following day in Chicago it may express a proposition judged to be false. Different times and places at which tokens are uttered will often (not always) determine in this way different referent occasions. This shift commonly occurs for sentences containing <u>indicator words</u>, words such as the demonstratives 'this' and 'that', pronouns such as 'he' or 'it', and 'here' and 'now', whose reference depends on context, and for sentences with tensed verbs. Logicians usually stipulate that what they mean by the term 'proposition' is what is expressed by a sentence relative to the referent occasion indicated by indicator words and verb tense and by the context in which a particular token of the sentence is used.[3] It follows that if what is expressed by different tokens of the same sentence is judged to have different truth values then different propositions are expressed. Thus, the utterance of 'It is raining' said in New York and the utterance of the same sentence in Chicago would express different propositions. To make explicit this difference we can specify the different referent occasions to form the different sentences, 'It is raining in New York on January 29, 1972' and 'It is raining in Chicago on January 30, 1972'. The two sentences can now be recognized as expressing different propositions without the need to refer to context.[4]

Acknowledging propositions as that to which truth and falsity are ascribed does not commit us, of course, to any particular metaphysical view about their nature. They may be identified with types of existent states of affairs; or necessary conditions for judgments of truth and falsity as psychological acts (the view this writer favors); or even ideal entities that we somehow apprehend, as for the Platonists. But a book on deductive logic is not the appropriate place to argue the merits of these alternatives. Our purposes are served by simply distinguishing propositions from sentences that express them.

4. Object Language and Meta-Language

There is still another distinction that is important for the study of logic, that between the language in which inferences are formulated and the language used to evaluate them as valid or invalid. The former is called the object language, the latter the meta-language, the language used by the logician in speaking about inferences formulated within the object language. The following inference, for example, is formulated in the object language:

> X must have killed Z, since either he or Y did, and we know Y is innocent

while the following sentence about the above inference is in the meta-language:

> 'X must have killed Z, since either he or Y did, and we know Y is innocent' is valid.

Here we refer to or mention the inference in the object language by enclosing it within single quotation marks and predicating of it the expression 'is valid'. In this manner the evaluation of the inference is formulated.

By this same device of quotation marks sentences of a meta-language can be formulated that mention also words and sentences of an object language. Thus, in the sentence

> Chicago is in Illinois

the word 'Chicago' occurs as part of a sentence of an object language, a sentence used here to convey information. On the other hand, in

> 'Chicago' contains seven letters

the quotation marks serve to name the word 'Chicago' in order to describe it, and the resulting sentence is formulated within a meta-language. Similarly, to say 'Chicago is in Illinois' is again to use the sentence of an object language, whereas to say '"Chicago is in Illinois" is true' is to mention the sentence and characterize it within a meta-language. The object language is thus a language in which words, sentences, and inferences are formulated and used for a variety of purposes in communication situations. The meta-language is a language in which sentences can be formulated that mention the words, sentences, and inferences of the object language by means of quotation mark names, and, most importantly for our purposes, is the language used by the logician in evaluating inferences.

This distinction of levels of language enables one to distinguish logic as a theory of inference from mathematics. The formulas of the various mathematical systems, e.g., 'The angles of a triangle are equal to 180 degrees' or '$x+y=y+x$', are formulas used in constructing these systems, in establishing axioms and deriving theorems, and should therefore be understood as formulated within an object language. So also are mathematical inferences, such as 'Since $x=y$, $z \cdot x = z \cdot y$'. On the other hand, such sentences as '"$x+y=y+x$" is true' and '"Since $x=y$, $z \cdot x = z \cdot y$" is valid' would be formulated in a meta-language, since they mention a formula and an inference by means of quotation marks in order to say something about them. They would represent the kind of sentences found in the logical evaluation of the formulas and inferences of mathematics. Logic and mathematics can be distinguished, then, by the levels of languages they both use. Mathematics is the science in which a certain type of symbolic language is used; logic is the theory of inference in which inferences formulated in this language are mentioned and evaluated, a theory which, as we shall presently see, also requires the use of symbols in order to perform this task.

The meta-language of logic can also, of course, be referred to by sentences of a language of a still higher level. Such would be the case in the sentence,

> '"Since $x=y$, $z \cdot x = z \cdot y$" is valid' is a sentence

in which the meta-linguistic sentence '"Since $x=y$, $z \cdot x = z \cdot y$" is valid' is in turn mentioned by means of quotation marks.

The sentence used in evaluating an inference is here mentioned in another sentence describing it. This sentence of the third level is a sentence of a meta-meta-language (a language about a language about an object language). And there seems no reason not to admit sentences that in turn mention these sentences within a fourth level, and so on indefinitely. For the purposes of the evaluation procedures we shall use, however, sentences of at most the third level prove to be sufficient.

Is it possible for a sentence to refer to itself, and thus allow us to collapse levels of language? In certain cases, at least, it seems that it is not. Consider, for example, the sentence,

1) This sentence is false.

No matter which truth value we assign to 1) we find ourselves in a contradiction. If 1) is false (expresses a false proposition), then it is false that the sentence is false, and hence 1) is true. While if 1) is true, then it is true that it is false, and hence 1) is false. Self-reference by sentences such as 1) is therefore clearly impossible, since it leads to contradictions.[1]

NOTES FOR CHAPTER I

The following notes provide references to supplementary material. Abbreviated titles of works are indicated in them. For the full titles see the bibliography at end.

Section 1.

1. See Toulmin [Argument] for a formulation of a general theory of inference and the distinction between deductive inferences and those of other types.

2. For a discussion of this form of inference in relation to Aristotle's "practical syllogisms" see Anscombe [Intention], pp. 57ff. See also von Wright [Practical Inference]. The term 'practical inference' as used here should not be applied to all inferences bearing on conduct, but only to those whose premisses describe or express a desire and belief. In particular, it should not be applied to the imperative and deontic inferences of Chapter VIII, though there is precedent for such usage (cf. Ross [Imperatives]). As noted below, these latter are to be regarded as deductive inferences.

3. The "conclusion" of this type of inference can be regarded as a tentative hypothesis and its "premisses" descriptions of the outcomes of empirical tests of the hypothesis. The hypothesis can be either a uniform generalization (of the form All S is P) or a statistical generalization (the probability of S being P is some rational number r). For introductions to this form of inference see Cohen and Nagel [Introduction], Bk II and the more recent Kyburg [Probability], Part II.

Section 3.

1. For reasons for distinguishing propositions from sentences see Frege [Thought]. More recently the role of the proposition has been assigned by Strawson to what he calls a "statement." See his [Logical Theory], pp. 3-5 and [Truth].
But there are a number of philosophers who have refused to make any distinction between sentences and what they express. For this view see especially Quine [Word], pp. 193-195. Quine seeks as a substitute for propositions what he calls "eternal sentences", sentences whose truth value does not change with circumstance of utterance. For criticisms of Quine's notion of eternal sentences see Cohen [Diversity], pp. 250ff.

2. For the necessity of distinguishing between the proposition (or statement) expressed by a sentence and its meaning see Lemmon [Sentences] and Pitcher [Truth], pp. 4-9.

3. Thus Frege in [Thought]: "But are there not thoughts [propositions] which are true today but false in six months time? The thought, for example, that the tree is covered with green leaves, will surely be false in six months time. No, for it is not the same thought at all. The words 'this tree is covered with green leaves' are not sufficient by themselves for the utterance, the time of utterance is involved as well. Without the time-indication this gives we have no complete thought, i.e. no thought at all. Only a sentence supplemented by a time-indication and complete in every respect expresses a thought." The same view with regard to statements is expressed by Strawson [Truth] and Lemmon [Sentences].

4. In Quine's terminology, to preserve constancy of truth value we convert the "occasion sentence" into two eternal sentences. See [Word], pp. 35-40, 193-195.

Section 4.

1. The paradox presented here is a version of what is referred to as the "liar's paradox." For a history of this paradox and a discussion of the modern solution in terms of language levels see the Kneales [Development]. It is one of a number of paradoxes which since Ramsay's [Foundations] (see pp. 1-61) have been distinguished as the "semantic paradoxes."

II. SENTENCE LOGIC

5. Logical Form

Classes of inferences can be distinguished according to the kind of connection that must hold between premisses and conclusion in order for a given inference to be valid. The most basic of such connections is between recurring propositions expressed by the constituent sentences of the premisses and conclusion. This connection defines the class of inferences evaluated in sentence logic.[1]

Let us adopt the convention of restating inferences by dividing premisses from conclusion by a straight line and representing the word 'therefore' by the symbol '∴'. Consider now the inference,

1) If it is cloudy, it will rain.
   It is cloudy.
   ∴ It will rain.

Notice that in it there are tokens of the sentence 'It will rain' in both the first premiss and the conclusion and of 'It is cloudy' in the two premisses. By means of these recurrences the conclusion is related to the premisses in such a way as to make it possible for the inference to be valid. It is not essential, however, that the same sentence recur in premisses and conclusion, as can be seen by the inference,

2) Either John hit Bill or Tom did.
   Bill was not hit by John.
   ∴ Tom hit Bill.

Here there are no recurring tokens of any one sentence. The same proposition, however, is expressed by the tokens of the different sentences 'John hit Bill' and 'Bill was hit by John', and also the same proposition is expressed by 'Tom did' and 'Tom hit Bill'. It is these recurrences of the same proposition expressed by different sentences that secures here the necessary connection between premisses and conclusion. It is propositions, then, and not sentences that must recur in the class of inferences we are considering.

Sentence logic may be defined, then, as that branch of logic in which inferences are evaluated whose validity depends on the recurrence of one or more propositions expressed by the constituent sentences of the premisses and conclusion. Typically at least one proposition expressed within the premisses is expressed also within the conclusion, as is the case for inferences 1) and 2). This enables the connection between premisses and conclusion upon which validity usually depends.

Compare now inference 1) to the inference,

    3) If these two lines are parallel, then they will never intersect.
    <u>These two lines are parallel.</u>
    ∴ They will never intersect.

It is evident that the two inferences are similar. In both a proposition expressed by the same sentence recurs in the first and second premisses and a second proposition recurs as expressed in the first premiss and conclusion. We can see this similarity more clearly if we use the letter A to represent the first sentence within the first premiss and B the second. Then inferences 1) and 3) can be regarded as of the general form,

    4) If A then B
        <u>A</u>
        B

We shall call representations such as 4) <u>inference forms</u>.

Both inferences 1) and 3) are valid. And it is clear that their validity does not depend on which sentences occur in the inferences, provided they recur in the same way and that the first premiss is of the form 'if...then...'. 4) thus represents a <u>valid inference form</u>, one that will represent a valid inference no matter which sentences are taken to be represented by the letters A and B. We might express this by saying that the inferences 1) and 3) are valid by virtue of their logical form represented by 4) and not by their content, by the particular sentences that happen to be the constituents of the inference. 4) represents this form by the letters A and B, the words 'if...then...', and the manner in which A and B recur within the inference. This feature of inferences 1) and 3) holds generally of the inferences evaluated within sentence logic. Their validity or invalidity does not depend on the particular propositions that recur, but on the form of the inference: the manner in which constituent sentences are combined by words such as 'if...then', 'and', and 'or', and in which sentences expressing the same proposition recur within the inference.

We shall use the capital letters from the first part of the alphabet A,B,C,... to represent arbitrary sentences of the object language. Thus, A could represent the sentence 'It is cloudy' of inference 1), 'These two lines are parallel' of 3), or any other sentence we may choose. These representing symbols are called <u>sentence schemata</u>. As in the inference form 4), these sentence schemata are used to represent the form of an inference. They are symbols of our logical meta-language used as a means of mentioning the constituent sentences of an inference in order to consider their truth or falsity and evaluate the inference as valid or invalid. It is arbitrary which of the schemata A,B,C,... we use to represent the sentences of an inference. The inference form 4) could have been written as

$$\frac{\text{If B then D}}{\text{D}}$$

with B representing 'It is cloudy' and D 'It will rain'. What is alone important is that the same sentence schema be used for sentences expressing the same proposition wherever they recur in the inference. If this is not done, the form of the inference, depending as it does on the nature of these recurrences, would not be represented.

Representing constituent sentences by sentence schemata is, as we have mentioned, not in itself sufficient to provide us with a means of evaluating a given inference as valid or invalid. Besides depending on the pattern of recurring sentences, validity also depends on definitions we give to the words that remain after schemata replace sentences. These will be <u>logical words</u> used to negate sentences or to combine them to form complex sentences, words like 'not', 'and', and 'if...then'. The inference form 4), for example, can be seen to be valid because we would define the words 'if...then' in such a way that if the first sentence within the first premiss expresses a true proposition then so must the second; since by the second premiss the first sentence is true, the conclusion follows. In order to evaluate more complex inferences we shall need exact definitions of these logical words and a procedure by which we can use the definitions to evaluate given inferences. The rest of this chapter is devoted to stating these definitions and developing the procedure for evaluation.

6. <u>Negation</u>, <u>Conjunction</u>, <u>and</u> <u>Disjunction</u>

<u>Negation</u>. Our primary use of the English word 'not' is to deny propositions that have been previously entertained or asserted. Thus, someone may say 'It is raining', and to indicate his disagreement another might reply 'No, it is not raining'. Or a scientist may tentatively propose the hypothesis 'Organism X has characteristic Y'. If this hypothesis conflicts with observed facts, this may lead him to deny it, to assert 'X does not have Y'. If we take the schema A to

represent 'It is raining' or 'X has Y', then by prefixing the symbol '-' representing the logical word 'not', we can form the schema -A representing 'It is not raining' or 'X does not have Y'. It is clear that from the use of 'not' that if the original proposition was indeed false, then its negation is true; while if the original proposition was justified as true, then the negation with which it has been denied is false. Thus, if 'It is raining' expresses a true proposition, 'It is not raining' or 'It is not the case that it is raining' expresses a false one; while if 'It is raining' is false, then 'It is not raining' is true. In general,

    i) If a sentence A is true (expresses a true proposition), -A is false.
   ii) If A is false, -A is true.

i) and ii) specify the conditions for the truth or falsity of -A, given the truth or falsity of A.

These conditions can be summarized in what is called a _truth table_ enumerating the possible determinations of truth and falsity of A and the corresponding determinations for -A.

A	-A
T	F
F	T

The 'T' in the column under A is to be understood as saying 'The sentence represented by A expresses a true proposition' and 'F' as 'The sentence A represents expresses a false proposition'. Since a proposition is either true or false, the truth table enumerates all the possible determinations of truth and falsity, or what are called _truth values_, of an arbitrary sentence A, or all its _possible interpretations_. Moreover, since a proposition cannot be both true and false at the same time, the possible interpretations of A exclude one another. There are thus two possible interpretations of A that are exhaustive (at least one of them must hold) and exclusive (no more than one can hold), and in terms of them the truth table specifies the truth values for its negation -A. Since the values for -A depend on those of A, -A is said to be a _truth function_ of A. The truth table provides a truth functional definition of logical negation by stating the conditions under which -A is true or false relative to the truth or falsity of A. This definition accords, as we have seen, with the use of 'not' in denying a proposition.

Note the contrast between the sign '-' used to represent the logical word 'not' or the phrase 'it is not the case that...' of the object language and the sentence schema A used to represent an arbitrary sentence. Unlike the sentence schema, the negation sign has a fixed content that will not vary with the sentence represented. We can use '-A', for example, to represent 'It is not raining', 'X does not have Y', and 'It is not the case that the 2 is rational'. The negation sign '-' represents the constant element in all these sentences while 'A' represents for that part that varies. While it is also used to represent the logical form of a sentence and inferences in which the sentence occurs, the manner of its representation is significantly different from that of the sentence schema. Signs standing for logical words like 'not' are called <u>logical constants</u>.

Conjunction. Our primary use of the English word 'and' is to combine sentences expressing propositions we wish to jointly assert or suppose as being true. Thus by means of 'and' we form a sentence such as 'It is raining and it is cloudy' from constituent sentences 'It is raining' and 'It is cloudy'. 'It is raining and it is cloudy' is called the <u>conjunction</u> of constituent sentences which are called the <u>conjuncts</u>. It is clear that the proposition expressed by the conjunction 'It is raining and it is cloudy' will be regarded as being true if and only if both conjuncts 'It is raining' and 'It is cloudy' are true. If one or both of them is false, then so is the conjunction formed from them. Since we have two constituent sentences and each can express a proposition either true or false, we have altogether four possible interpretations of the sentences, and these possibilities will determine the possible truth or falsity of the sentence formed from them by means of 'and'. Using A and B as our sentence schemata and the <u>conjunction sign</u> '∧' to stand for 'and', then we can form the schema A∧B to represent the conjunction 'It is raining and it is cloudy'. The truth values for A∧B for the four possible interpretations of its conjuncts are then

    i) If A and B are both true, so is A∧B.
    ii) If A is true and B false, then A∧B is false.
    iii) If A is false and B true, then A∧B is false.
    iv) If A and B are both false, then A∧B is false.

These conditions for the truth and falsity of A∧B can again be summarized in a truth table.

A	B	A∧B
T	T	T
T	F	F
F	T	F
F	F	F

These four possible interpretations of A and B are again exhaustive and exclusive. The specification of the conjunction $A \wedge B$ for each of them constitutes our truth functional definition of the conjunction sign. Because the words it is used to represent connect or combine sentences the conjunction sign, like the other logical constants to be introduced in this chapter, is called a <u>logical connective</u>. For the sake of simplicity the negation sign is usually also listed as one of the sentence connectives, though it only operates on and never connects schemata.

The conjunction sign '$\wedge$' can be used to represent a variety of words in the English language other than 'and'. The sentences 'Tom played well, but he was sick', 'Tom played well, although he was sick', 'Tom played well; nevertheless, he was sick', and 'Tom played well because he was sick' would all be represented by $A \wedge B$. There are, to be sure, differences in meaning between 'and', 'but', 'although', 'nevertheless', and 'because'. However, the manner in which the truth or falsity of these sentences formed by means of them depends on the truth or falsity of 'Tom played well' and 'he was sick' is the same. Since this is what alone is important to the logician in representing the form of these sentences, he can ignore the differences in meaning when employing the conjunction sign.

<u>Disjunction</u>. We also combine sentences by means of the word 'or', as in such sentences as 'It is raining or it is snowing' and 'Either I will go to the movie or I will study'. Such sentences are called <u>disjunctions</u> and their constituent sentences <u>disjuncts</u>. The word 'or' is used in English with two senses. In one sense a disjunction is true if one or the other of its disjuncts are true, but false if both are true or both are false. In 'Either I will go to the movie or I will study' the expression 'either...or' seems to be used in this sense; there seems to be the implication on the part of the speaker that he will not both go to the movie and study and if he should do both we would judge his statement false. This is called the <u>exclusive</u> sense of 'or'. In a second sense called the <u>inclusive</u> sense a disjunction is regarded as true if both disjuncts are true as well as if just one is. To make this sense explicit we often write the expression 'and/or'. Our logical definition of 'or' must choose between these alternatives. It proves to be more convenient in evaluating inferences to choose the inclusive sense, which we represent by the disjunction sign '$\vee$'. The truth table for this logical connective is accordingly

A	B	A∨B
T	T	T
T	F	T
F	T	T
F	F	F

    We can, of course, form conjunctions and disjunctions from negated sentences, as in 'It is raining and it is not cloudy' and 'Either it is not raining or it is not snowing'. These would be represented by A∧-B and -A∨-B. We can also negate a conjunction or a disjunction, as in 'It is not the case that it is raining and it is cloudy'. But here we face an ambiguity that we must resolve. For this could be understood as the conjunction of the negated sentence 'It is not raining' with 'It is cloudy' or the negation of a conjunction. To resolve this ambiguity we must introduce into our symbolism means of indicating the intended grouping of sentences called <u>parentheses</u>. The conjunction of a negation with another sentence we would represent by -A∧B, with the negation sign operating on A. The negation of a conjunction, on the other hand, would be represented as -(A∧B), with the parentheses indicating the intended grouping of sentences. Parentheses have a similar use when we connect disjunctions with conjunctions or conjunctions with disjunctions, as in 'It is raining and it is cloudy or it is snowing'. This could be understood as either the conjunction of 'It is raining' with the disjunction 'It is cloudy or it is snowing' or as the disjunction of 'It is snowing' with a preceding conjunction. If C be taken as representing 'It is snowing', then the former can be represented as A∧(B∨C), the latter as (A∧B)∨C, the parentheses again distinguishing the intended groupings. Brackets and braces may be introduced for the representation of more complex sentences. Thus, 'It is not the case that it is either raining and cloudy or it is snowing' would be represented by -[(A∧B)∨C], with the pair of brackets enclosing the schema that is negated. By introducing braces we may form a schema such as -{[(A∧B)∨C]∧D}, with braces now enclosing the negated schema.

    By negating sentences by means of 'not' and combining them by 'and' and 'or' successively more complex sentences can be generated. The sentences that are negated or combined, e.g. 'It is raining', 'It is cloudy', are called <u>atomic sentences</u>; those generated by negation and combination, e.g. 'It is raining and it is either not cloudy or it is snowing', are called <u>molecular sentences</u>. These sentences are represented by <u>atomic</u> and <u>molecular sentence schemata</u> respectively.

    Using our definitions of negation, conjunction, and disjunction we can determine by means of a truth table the truth values of molecular sentences for all possible interpretations of their atomic constituents. Our procedure here is to use the definitions in determining successively larger constituents of the molecular sentence. Thus, for the sentence, 'It is raining and it is either not cloudy or it is snowing', whose logical form is A ∧ (-B ∨ C), we would determine first the truth values of -B relative to the possible interpretations of B, then the values of -B ∨ C relative to the possible interpretations of B and C, and then finally the values for the whole molecular schema. Since there are three atomic sentences within this sentence, each expressing a proposition that is true or false, there are eight possible interpretations of the atomic constituents. The truth table for our example takes the following form:

A	B	C	-B	-B ∨ C	A ∧ (-B ∨ C)
T	T	T	F	T	T
T	T	F	F	F	F
T	F	T	T	T	T
T	F	F	T	T	T
F	T	T	F	T	F
F	T	F	F	F	F
F	F	T	T	T	F
F	F	F	T	T	F

From the truth table we can see that the sentence is true if it is raining and cloudy and snowing or if it is raining and not cloudy and snowing or if it is raining and not cloudy and not snowing. For all other possible interpretations of the atomic constituents the sentence expresses a false proposition. In general, for n constituent sentences there will be 2^n possible interpretations.

The above truth table can be greatly simplified by writing the truth value for a given schema under the logical connective that forms it. The truth values for the whole molecular schema then become those under the connective operating on or combining the largest constituents, or what is called the <u>main connective</u>. Adopting this procedure, our truth table becomes

A	B	C	A	∧	(-B	∨	C)
T	T	T	T	T	F	T	T
T	T	F	T	F	F	F	F
T	F	T	T	T	T	T	T
T	F	F	T	T	T	T	F
F	T	T	F	F	F	T	T
F	T	F	F	F	F	F	F
F	F	T	F	F	T	T	T
F	F	F	F	F	T	T	F

Since '∧' is the main connective, the values under it are the values of the whole molecular schema. This is indicated by lines enclosing these values.

Exercises

I. Represent the logical form of the following sentences. As an aid in their representation use B to represent 'Bill went to the store', C 'Charlie went to the store', and R 'It rained yesterday'.
1. Bill and Charlie went to the store.
2. It is not the case that either Bill or Charlie went to the store.
3. Either Bill didn't go to the store or Charlie did.
4. Either Bill and Charlie went to the store or it rained yesterday.
5. Either Bill went to the store or Charlie went and it rained yesterday.
6. Bill went to the store, although Charlie went also and it rained yesterday.
7. Bill went to the store, but neither did Charlie go nor did it rain yesterday.
8. Bill didn't go to the store because it is not the case that either Charlie went or that it rained yesterday.
9. Neither Bill nor Charlie went to the store, and yet it rained yesterday.
10. It is not the case that either Bill or Charlie went to the store without it raining yesterday.

II. Assume that it is true that Bill went to the store, true also that Charlie went, but false that it rained yesterday. Determine the truth values of the propositions expressed by the above sentences given these assumptions.

III. Construct truth tables listing truth values relative to the possible interpretations of their constituents for sentences 2-6 of I.

7. Implication, Equivalence, and the Stroke Function

Material Implication. A conditional is any molecular sentence of the form 'if A then B' or 'B provided that A', in which one sentence, the antecedent of the conditional, is stated to be a condition for the truth of another, the consequent. There are a variety of types of conditionals. An especially important one of these for us is the logical conditional in which it is asserted that the conclusion of an inference follows from the premises. Any inference with two premises P_1 and P_2 and conclusion C has as its standard form,

$$\frac{\begin{array}{c}P_1\\P_2\end{array}}{C}$$

But we can represent the inference also by a single sentence of the form,

If P_1 and P_2 then C.

Here 'if...then' is used in the sense of 'logically implies', the claim being made that the inference represented is valid, that it is impossible for P_1 and P_2 be true and C false.

A second important type of conditional is the <u>counterfactual conditional</u>, a conditional with false constituents used to assert that a certain connection holds between natural events. On the basis of a number of observations we may conclude, for example, that whenever sugar is placed in water it will dissolve. And from this generalization we could infer the counterfactual conditional, 'If this piece of sugar had been placed in water, then it would have dissolved'. More generally, from any law-like generalization of the form, 'Whenever a type of event C occurs, then so does event E' can be inferred a counterfactual of the form 'If a particular event of type C were to occur, then a particular E would also'. This latter sentence in effect asserts that a general regularity in nature can be extended to events that have not yet taken place. Both the antecedent and the consequent of the counterfactual are false: neither the particular event C (the placing of this sugar in water) nor the event E (its dissolving) have in fact occurred. What we assert in this type of conditional is that nature is so regulated that <u>if</u> the antecedent <u>were</u> to be true so <u>would</u> the consequent.

There are other types of conditionals as well, those represented, for example, by 'If it is cloudy, it will rain', 'If he comes, I will greet him', etc. Despite important differences between the logical and counterfactual conditionals and between them and a variety of others, all seem to have at least this basic sense: if the antecedent is true and the consequent false, then the conditional as a whole is false. Thus, true premisses and a false conclusion falsify a logical conditional or result in an invalid inference; for the event C to occur and E not would falsify the generalization that whenever C occurs E does also, and hence falsify also the counterfactual conditional which it supports; and similarly for the others. This common sense is the basis for the truth table definition of the logical connective '⊃' used to represent the expression 'if...then'. With A as the antecedent and B as the consequent this definition is as follows:

A	B	A⊃B
T	T	T
T	F	F
F	T	T
F	F	T

The conditional defined by this truth table is called <u>material implication</u> and the logical connective combining the antecedent and consequent the <u>material implication sign</u>.

Note that the material implication sign just defined is much weaker than the 'if...then' of the logical and counterfactual conditionals, agreeing with their sense only in the second row of the truth table. For whereas A⊃B is true if A and B are, the truth of the premisses and conclusion does not guarantee a valid inference. ('It is raining' may be true and so may 'It is raining and cloudy', but the latter cannot be inferred from the former.) Again, the third row of the truth table states that A⊃B is to be taken as true if A is false and B is true. But this definition cannot be applied to the logical conditional. ('It is raining' may be false and 'It is cloudy' true, but there is no valid inference from the former to the latter.) And finally, in the fourth row the material implication is true when both its constituents are false. Yet this does not accord with the sense of the counterfactual conditional, for it may be false that I put this piece of iron in water and false that it dissolves, but the counterfactual 'If this iron were to be put in water, it would dissolve' is certainly not true. Only in the second of the possible interpretations of its constituents, therefore, does the material implication correctly reflect all types of conditionals.

Why then is there the value 'true' in the first, third, and fourth interpretations? The answer seems to be that this value is dictated by the basic assumption that the proposition expressed by a sentence is either true or false. We would certainly not want to say that if the antecedent and consequent are true, then the conditional is false. But since it is not false and a proposition must be either true or false, we are lead to list it as true. A false antecedent and true consequent also does not necessarily give us a false conditional, for an inference can be valid with a false premiss and true conclusion. Nor should false antecedents and consequents give false conditionals, since true counterfactual conditionals have such constituents. But if we cannot supply 'false' for these last two interpretations, our logic requires again the value 'true'. 'True' in the definition of the material implication thus means 'not false'. Our definition of the material implication given in the truth table should be thus understood as stating that the implication is false if the antecedent is true and the consequent false; otherwise it is not false, and since there is only one alternative, it is therefore true. Though the implication sign just defined does not accurately represent any of the principal uses of the expression 'if...then', it does prove sufficient to allow the evaluation of inferences in which conditionals occur as premises or conclusions. Hence its adoption by the logician.

But the consequences of this definition appear to clash with our intuitions. It follows from the definition that a material implication is true if its antecedent is false, regardless of what consequent occurs. Also, a material implication is true if its consequent is true, regardless of the antecedent. Hence the two sentences,

      1) If 2 + 2 = 5, then the moon is made of green cheese
and 2) If the moon is made of green cheese, then 2 + 2 = 4

are both regarded as true. These two consequences are known as the "paradoxes of material implication." But they are paradoxical only if we mistakenly identify a material implication with either a logical conditional describing a logical connection or a counterfactual conditional describing a causal connection in nature and expect that every 'if...then' sentence must correctly describe such connections in order to be true. There is, to be sure, no connection of any kind between the antecedents and consequents of 1) and 2). They are nevertheless true if 'if...then' is interpreted in the very restricted sense of a material implication that is a truth function of its antecedent and consequent.[1]

As before, the expressions 'if...then' can be used together with the other logical words to generate increasingly complex molecular sentences from atomic sentences, e.g. 'If the Army lowers its expenditures and so does the Navy, then either we will not have inflation or there will be unemployment'. With parentheses again indicating groupings of sentences this sentence could be represented by the schema $(A \wedge B) \supset (-C \vee D)$.

Material Equivalence. A biconditional is any sentence of the form 'If A then B and if B then A' or 'A if and only if B'. Again there are many types of these sentences, the most important being the logical and counterfactual biconditionals. And again the logician adopts a form that is weaker than any of them. It is called <u>material equivalence</u> and has the following truth table:

A	B	A≡B
T	T	T
T	F	F
F	T	F
F	F	T

A material equivalence is thus true when its constituents have the same truth value; otherwise it is false.

The truth table for $A \equiv B$ is the same as for $(A \supset B) \wedge (B \supset A)$. Since $B \supset A$ is logically equivalent to $-A \supset -B$, $A \equiv B$ can be understood as $(A \supset B) \wedge (-A \supset -B)$. When $A \supset B$ is true, A is said to be the <u>sufficient</u> condition for B (if it is true, so is B). When $-A \supset -B$ is true, A is said to be the <u>necessary</u> condition for B (if it is false, so is B, or B is true only if A is also). The material equivalence $A \equiv B$ can thus be understood as stating that A is the <u>necessary</u> and <u>sufficient</u> condition for B.

The material implication and equivalence signs can be used to represent the frequently used word 'unless'. When we assert a sentence of the form 'A unless B' (e.g. 'It will get hot unless it rains') we seem to be making two assertions: first, that if B is true (it rains), then A will be false (it will not get hot), that is, $B \supset -A$, or alternatively $A \supset -B$; and secondly, that if B is false (it does not rain), then A will be true (it will get hot), or $-B \supset A$. 'A unless B' can thus be represented by the conjunction $(A \supset -B) \wedge (-B \supset A)$, and thus by definition of the material equivalence sign by $A \equiv -B$. A sentence of the form 'Not A unless B' (e.g. 'It won't cool off unless it rains'), on the other hand, would seem to have the same meaning as 'A only if B' or 'B is a necessary condition for A', and thus should be represented by $-B \supset -A$ or $A \supset B$.

To reduce the need of parentheses in increasingly complex schemata that can be generated by the connectives introduced let us adopt the binding convention that states that the negation sign has stronger binding force than the conjunction sign (a convention we have been already using), and that the conjunction, disjunction, material implication, and material equivalence signs have progressively weaker binding forces. Using parentheses, brackets, and braces as grouping devices, we may form by our connectives such a schema as $\{[(-A \wedge B) \vee C] \supset (B \wedge D)\} \equiv (A \vee C)$. Since our binding convention states that '$\wedge$' has a stronger binding force than '$\vee$', $(-A \wedge B) \vee C$ can be written as simply $-A \wedge B \vee C$, the convention indicating the intended grouping. Similarly, since both '$\wedge$' and '$\vee$' have stronger binding force than '$\supset$', $[-A \wedge B \vee C] \supset (B \wedge D)$ may be written as $-A \wedge B \vee C \supset B \wedge D$. Finally, using the convention that '$\equiv$' has the weakest binding force, the whole schema may be written without any parentheses as $-A \wedge B \vee C \supset B \wedge D \equiv A \vee C$. It should not be thought, however, that the binding convention just adopted allows us to dispense altogether with parentheses. The schema $(A \vee B) \wedge (C \supset D)$, for example, must have the parentheses indicated. Without them the binding convention would dictate an entirely different grouping. Another convention regarding parentheses also proves useful. Since a conjunction $A \wedge (B \wedge C)$ has the same truth table interpretation as $(A \wedge B) \wedge C$, the order of parentheses is irrelevant.

They can therefore be dropped and the conjunction written as $A \wedge B \wedge C$. For a conjunction with n conjuncts $A_1, A_2, \ldots, A_n$ we can thus write $A_1 \wedge A_2 \wedge \ldots \wedge A_n$. Similarly for a disjunction of n disjuncts we may drop parentheses and write $A_1 \vee A_2 \vee \ldots \vee A_n$.

The Stroke Function. Because the truth table for $A \equiv B$ is the same as $(A \supset B) \wedge (B \supset A)$ we could regard the former as definable by the latter and introduce the following definition:

$$A \equiv B =_{df} (A \supset B) \wedge (B \supset A)$$

Here the definiendum $A \equiv B$ may be regarded as being introduced as a short-hand expression for the definiens, a means of abbreviating what is otherwise cumbersome. It is also possible to define any of the connectives in terms of the negation sign and one other connective. $A \supset B$, for example, may be defined in terms of negation and conjunction as $-(A \wedge -B)$ or in terms of negation and disjunction as $-A \vee B$; $A \vee B$ may be defined in terms of negation and conjunction as $-(-A \wedge -B)$ or in terms of negation and implication as $-A \supset B$; and similarly for the conjunction sign. We may refer to such definitions as the reduction of one connective to the negation sign and some other connective.

It is possible to carry this reduction process still further and define all the logical connectives so far introduced, including the negation sign, with just one connective. Since for two atomic schemata A and B there will be four possible interpretations, and since a two-place connective is defined as a selection of 'true' or 'false' for all of the four interpretations, there are altogether sixteen possible connectives that could be defined. Two of these sixteen possible connectives (most not corresponding to any logical words in English) are suitable for defining the logical connectives we have introduced. The most commonly used of these is represented by the stroke connective '|', and has the following definition:

A	B	A|B
T	T	F
T	F	T
F	T	T
F	F	T

$A|B$ may be understood as representing the incompatibility or mutual exclusiveness of A and B, since $A|B$ is false when both A and B are true, and otherwise true. It can be easily verified by the reader that $-A$ can be defined by $A|A$ and $A \wedge B$ by $(A|B)|(A|B)$. Since the remaining three connectives are definable in terms of negation and conjunction, they can be similarly reduced to the stroke connective.

Exercises

I. Represent the logical form of the following sentences, using B to represent 'Bill went to the store', C 'Charlie went to the store', R 'It rained yesterday', and S 'It snowed yesterday'.

1. If Bill went to the store, then Charlie did also and it rained yesterday.
2. If Bill and Charlie went to the store, then it didn't snow yesterday.
3. If Bill went to the store then if Charlie went also it rained yesterday.
4. If Bill's going to the store implies Charlie went also, then it rained yesterday.
5. Charlie went to the store only if it rained yesterday.
6. Charlie went to the store only if it neither rained nor snowed yesterday.
7. Charlie went to the store provided it either didn't rain or didn't snow.
8. Unless it rained yesterday Bill or Charlie went to the store.
9. Bill went to the store unless it either rained or snowed yesterday.
10. Bill didn't go to the store unless it either rained or snowed yesterday.
11. Bill and Charlie went to the store only if it neither rained nor snowed yesterday.
12. Neither Bill nor Charlie went to the store unless it is not the case that it either rained or snowed yesterday.
13. If Bill went to the store, then Charlie went also unless it rained yesterday.
14. Bill and Charlie went to the store if and only if it didn't rain yesterday.
15. That Bill and Charlie went to the store is necessary and sufficient for it not snowing yesterday if it rained.

II. Assume that it is true that Bill went to the store, true also that Charlie went, but false both that it rained yesterday and that it snowed. Determine then the truth values of the above sentences.

III. Construct truth tables for sentences 1-5 of I.

8. Evaluation of Inferences

We are now at the gateway to the goal of our endeavors of this chapter, the evaluation of inferences whose validity depends on the relation between propositions expressed by constituent sentences within the premisses and conclusion. Consider the inference,

1) If Alfred had committed the murder, he would have been absent from the party.
But he was not absent from the party.
∴ Alfred did not commit the murder.

This would be valid if it were impossible for the premisses to be true and the conclusion false. That this is indeed the case can be seen by determining the truth values of the premisses and conclusion for the possible interpretations of the atomic sentences 'Alfred committed the murder' and 'Alfred was absent from the party' and seeing whether for any of these interpretations both premisses are true and the conclusion false. The logical representation of 1) is

2) $A \supset B$
$-B$
$\overline{-A}$

The truth table is constructed by determining the truth values for the premisses $A \supset B$ and $-B$ and then the conclusion $-A$ relative to the possible interpretations of the atomic constituents A and B.

A	B	$A \supset B$	$-B$	$-A$
T	T	T	F	F
T	F	F	T	F
F	T	T	F	T
F	F	T	T	T

Since there is no interpretation in which both premisses are true and the conclusion false, any inference of the form 2) is valid, and hence 1) is shown to be valid. The truth of its premisses would guarantee the truth of its conclusion that Alfred was innocent.

In general, any inference can be evaluated by representing its logical form and then constructing a truth table listing the possible interpretations of its atomic constituents. If for all the 2^n possible interpretations of n atomic constituents there is none in which the premisses are all true and the conclusion false, the inference is valid. Otherwise, the inference is invalid.

The evaluation procedure just outlined can be applied without difficulty to inferences like 1) whose constituent sentences can be readily identified and logical form easily represented. To be sure, for any inference with a large number of constituent sentences the truth table that is to be constructed will contain a large number of possible interpretations and the calculation of truth values for the premisses and conclusion will be complex. But these are mechanical difficulties that can be overcome with time, patience, and attention to detail. Very seldom in actual conversation or writing, however, do we encounter inferences with such perspicuous logical form. Especially with inferences stated with argument force (cf. Section 2), logical form can be disguised in a variety of ways, and it may be very difficult to single out the constituents of the inference.

Consider, for example, Aquinas' well-known "argument from design" or "fifth way" in his <u>Summa Theologica</u>.

> The fifth way is taken from the governance of the world. We see that things which lack knowledge, such as natural bodies, act for an end, and this is evident from their acting always, or nearly always, in the same way, so as to obtain the best result. Hence it is plain that they achieve their end, not fortuitously, but designedly. Now whatever lacks knowledge cannot move towards an end, unless it be directed by some being endowed with knowledge and intelligence; as the arrow is directed by the archer. Therefore some intelligent being exists by whom all natural things are directed to their end; and this being we call God.

The word 'therefore' in the last sentence plainly separates off the conclusion of the inference from the premisses. But how shall we represent the logical form of this inference?

To answer this we must note, first of all, that since this is an argument, Aquinas is concerned to establish the truth of the premisses, and therefore adds justifications or illustrations for them that are irrelevant to determining the validity of the inference. The second sentence of the argument, 'We see that...as to obtain the best result' and the clause at the end of the next to last sentence, 'as the arrow is directed by the archer', are such supplementing material that can be deleted without altering the deductive relation between premisses and conclusion. Without them and the first sentence we are left with

> ...it is plain that they [natural bodies] achieve their end, not fortuitously, but designedly.
> Now whatever lacks knowledge cannot move towards an end, unless it be directed by some being endowed with knowledge and intelligence.
> ∴ Some intelligent being exists by whom all natural things are directed to their end; and this being we call God.

There must be sentences expressing the same propositions in the premisses and the conclusion in order for this to be a valid inference (cf. Section 5). In this inference there are different linguistic formulations of two different propositions. If we use the same sentence to express the same proposition, then our inference may be reformulated as

> Natural bodies achieve their end by design.
> Unless there exists the being endowed with intelligence whom we call God to direct them, natural bodies would not achieve their end by design.
> ∴ There exists the being endowed with intelligence whom we call God to direct them.

Representing the constituent sentences now by A and B, the logical form of this inference can easily be seen to be

$$\frac{A}{-B \supset -A}$$
$$B$$

If we were to construct a truth table, we would find it is impossible for premisses of the form B and $-A \supset -B$ to be true and A false. Aquinas' inference is thus valid. It does not, of course, follow that we must accept his further claim that the conclusion is true, that the argument is sound (cf. Section 2); we may still remain unconvinced of the truth of one or both of the argument's premisses.

There are still other ways in which logical form can be disguised. Consider the inference,

> The butler must have murdered Smith, since the chauffeur didn't.

As is common for inferences with argument force, the conclusion is stated first instead of last as required for our standard form. Also, there is an implied premiss which must be supplied if the conclusion is to be established. Supplying this premiss, reordering the premisses and conclusion, and expressing the same proposition by the same sentence, we have

> The chauffeur did not murder Smith.
> Either the butler or the chauffeur murdered Smith.
> ∴ The butler murdered Smith.

This can now be easily represented by sentence schemata and evaluated. Inferences such as this with an implied premiss are called <u>enthymemes</u>.

There are, then, altogether three ways that logical form can be disguised: i) there may be supporting or illustrative material within the inference irrelevant to the connection between premisses and conclusion; ii) the same proposition may be expressed by different sentences; iii) the conclusion may precede one or more of the premisses; iv) premisses may be missing. Evaluating an inference requires first uncovering these "disguises" and then representing the inference's form with sentence schemata. This uncovering of logical form can be the most difficult step in evaluation, requiring often considerable ingenuity in contrast to the mechanical procedures that can be applied when it is completed.

Exercises

I. Use truth tables to determine whether or not the following inference forms are valid or invalid.

1. $\dfrac{A \land B}{A}$ 6. $\dfrac{B}{A \supset B}$ 11. $\dfrac{A \supset B}{-A \supset -B}$

2. $\dfrac{A}{A \land B}$ 7. $\dfrac{-A}{A \supset B}$ 12. $\dfrac{A \supset (A \supset B)}{B \supset (A \supset B)}$

3. $\dfrac{A \lor B}{A}$ 8. $\dfrac{A \supset B,\; B}{A}$ 13. $\dfrac{A,\; -A}{B}$

4. $\dfrac{A}{A \lor B}$ 9. $\dfrac{A \supset B,\; -A}{-B}$ 14. $\dfrac{A \equiv (B \equiv A)}{B}$

5. $\dfrac{A}{A \supset B}$ 10. $\dfrac{A \supset B}{-B \supset -A}$ 15. $\dfrac{A \land (A \supset B),\; (A \lor B) \equiv B}{-B \equiv A}$

II. Represent the logical form of the following inferences and evaluate them as valid or invalid by means of truth tables. Choose sentence schemata that will facilitate representation of logical form. For inference 6, for example, D might be chosen to represent 'The defendant murdered Smith', S to represent 'He was at the scene of the crime', and J to represent 'Jones would have seen him'.

   1. If Bill went to the store, then if Charlie did also it rained yesterday. But it didn't rain yesterday. Therefore, either Charlie didn't go to the store or Bill did.

   2. If Bill went to the store, then it either rained or snowed yesterday. But it neither rained nor snowed yesterday. Therefore, Bill didn't go to the store.

3. If Bill went to the store, then if it rained yesterday Charlie also went. Therefore, if it rained yesterday, then if Charlie didn't go to the store neither did Bill.

4. Bill didn't go to the store yesterday unless it rained and Charlie went also. Charlie did go to the store, although it rained. Therefore, Bill went to the store.

5. Only if it rained yesterday did either Bill or Charlie go to the store. Both Bill and Charlie went to the store. Therefore, it must not have rained.

6. If the defendant murdered Smith then he was at the scene of the crime. If he was at the scene of the crime, then Jones would have seen him. Now Jones did see him. Therefore, the defendant murdered Smith.

7. If the defendant murdered Smith then he was at the scene of the crime and Jones saw him. Now Jones didn't see the defendant. Therefore, the defendant didn't murder Smith.

8. The square root of 2 is either rational or irrational. If it is rational the hypotenuse of a right-angled triangle is divisible into equal parts. But the hypotenuse of a right-angled triange is not divisible into equal parts. Therefore, the square root of 2 is irrational.

9. This triangle is either isosceles or equilateral. If it is equilateral its angles are equal. Therefore, either its angles are equal or it is isosceles.

10. If Newton's theory of gravitation is true then the velocity of a particle increases if and only if its mass does not increase. We can observe that the mass of a particle does increase provided its velocity increases. Hence, Newton's theory is mistaken and Einstein's theory is correct.

11. If this specimen were excessively heated it would be destroyed. If it were examined by means of an electron microscope it would be excessively heated. Therefore, the specimen will be destroyed provided it is examined by an electron microscope.

12. God is by definition a being greater than which nothing can be conceived. If God did not exist then we could conceive of a being greater than Him, namely one who had all His attributes plus that of existence. Hence, God must exist.

13. Descartes' epistemology is mistaken. For if it were correct, then we could never justifiably claim any empirical proposition as certain. If we could never claim any empirical proposition as certain, the word 'certain' could have no use within descriptive discourse. And, in fact, there is an accepted use of 'certain'.

14. "...nor can any one apply them (words) as marks, immediately, to anything else but the ideas that he himself hath: for this would be to make them signs of his own conceptions and yet apply them to other ideas; which would be to make them signs and not signs of his ideas at the same time;..." (Locke's *Essay Concerning Human Understanding*, Book II, Chapter II, Section 2)

15. "Philonous: Suppose now one of your hands hot, and the other cold, and that they are both at once put into the same vessel of water, in an intermediate state; will not the water seem cold to one hand, and warm to the other?
"Hylas: It will.
"Philonous: Ought we not therefore, by your principles, to conclude it is really both cold and warm at the same time, that is, according to your own concession, to believe an absurdity.
"Hylas: I confess it seems so.
"Philonous: Consequently, the principles themselves must be false, since you have granted that no true principle leads to an absurdity."
(Berkeley's <u>Three Dialogues</u>, Dialogue I)

9. <u>General Decision Procedure</u>

So far we have restricted ourselves to constructing a procedure which would enable us to decide for any given inference of a certain kind whether or not it is valid. We can extend this procedure to one that enables us to decide whether or not given molecular sentences can possibly be false. This extension is called the <u>general decision procedure</u> for sentence logic.

Consider, for example, the sentence,

1) It is raining or it is not raining

of the form $A \vee -A$. That this sentence cannot possibly be false (or alternatively, is necessarily true) is shown by the following truth table:

A	$A \vee -A$
T	T F
F	T T

For the two possible interpretations of its one atomic constituent A, $A \vee -A$ is shown to be true. No matter whether or not it is, in fact, raining sentence 1) will thus be true. Its truth is independent of what we observe the facts to be and dependent only on its logical form. Any other sentence of this form will be true for exactly the same reason. The propositions expressed by sentences such as 1) whose truth depends only on logical form as shown by a truth table decision procedure are called <u>tautologies</u>.

There is a second class of propositions that our truth table decision procedure enables us to distinguish from others. These are <u>contradictions</u>, propositions that cannot possibly be true. An example of a contradiction would be the proposition expressed by

2) It is raining and not raining

of the form A∧-A. That this is necessarily false is again disclosed in a truth table.

A	A∧-A
T	F F
F	F T

No matter whether it is true that it is raining or false, 2) must be judged false because of its logical form. Still a third class of propositions are the <u>contingent propositions</u> that are neither tautologies nor contradictions. These are propositions such as that expressed by 'It is raining or it is cloudy' which are true under some interpretations of their constituents but false under others. Those sentences we use to convey information including the generalizations of science all express contingent propositions.

It is evident that every proposition will be either a tautology, a contingent proposition, or a contradiction. The decision procedure for sentence logic enables us to determine for any arbitrary sentence to which class the proposition it expresses belongs. If for all 2^n possible interpretations of its n atomic constituents the proposition it expresses is true, then the sentence expresses a tautology; if it is true for some of these interpretations and false for others, it expresses a contingent proposition; if it is false under all interpretations, it expresses a contradiction.[1]

Our evaluation procedure for inferences developed in the previous section can be regarded as but a special case of the decision procedure just outlined. This is due to the fact noted at the beginning of Section 7 that every inference can be put in the form of a conditional sentence in which the conjunction of the premises forms the antecedent and the conclusion the consequent. For example, an inference such as 'Either the butler or the chauffeur murdered Smith. The chauffeur did not murder him. ∴ The butler murdered Smith' can be put in the form of the conditional,

3) If either the butler or the chauffeur murdered Smith and the chauffeur did not murder him then the butler murdered Smith.

It is clear that whether or not the inference in question is valid can be decided by determining whether 3) expresses a tautology, since it will then be impossible for the antecedent of the conditional (the premises of the inference) to be true and the consequent (the conclusion) false. The form of 3) is (A∨B)∧-B⊃A. Its truth table reveals that, in fact, it does express a tautology.

A	B	(A∨B)	∧	-B	⊃	A
T	T	T	F	F	T	T
T	F	T	T	T	T	T
F	T	T	F	F	T	F
F	F	F	F	T	T	F

Hence 3) is valid. In general, any inference with k number of premises of the form $P_1, P_2, \ldots, P_k \therefore C$ can be put in the form of a conditional $P_1 \wedge P_2 \wedge \ldots \wedge P_k \supset C$. If this conditional is true for all 2^n possible interpretations of the atomic constituents of $P_1, P_2, \ldots, P_k, C$, then the inference is valid; otherwise it is invalid.

It is useful to employ a special term for the relation that holds between the antecedent and consequent of a conditional that is necessarily true. We call this the relation of <u>entailment</u> and symbolize it by '$\Rightarrow$'. Since we have shown that a sentence of the form $(A \vee B) \wedge -B \supset A$ expresses a necessary truth, we have established that the entailment relation $(A \vee B) \wedge -B \Rightarrow A$ holds between the antecedent and consequent. In general, the entailment relation holds between the premises and conclusion of every valid inference, or if an inference $P_1, P_2, \ldots, P_k \therefore C$ is valid, then $P_1 \wedge P_2 \wedge \ldots \wedge P_k \Rightarrow C$. It is important to distinguish the symbol '$\Rightarrow$' representing a logical <u>relation</u> holding between propositions from the material <u>implication</u> sign '$\supset$' as a logical <u>connective</u>. The latter is used to represent conditionals of the <u>object</u> language preparatory to our evaluating them. But we use the entailment symbol to represent the sentence within our meta-language reporting the results of this evaluation. Thus, the schema $A \wedge B \supset A$ might represent the object language sentence 'If it is both raining and cloudy then it is raining'. In contrast, the expression $A \wedge B \Rightarrow A$ would represent the sentence within the meta-language '"It is both raining and cloudy" entails "It is raining"' in which sentences are mentioned and the result of an evaluation procedure reported.[2]

Notice that in the expression $A \wedge B \Rightarrow A$ the schemata stand in place of <u>names</u> of sentences, and not the sentences themselves. As <u>they</u> occur in the truth tables they also have this function, for when we interpret a constituent schema A as true within a truth table we are predicating within the meta-language the adjective 'true' of the name of some arbitrary sentence we are taking it to represent (cf. Section 4). Thus, if A represents the sentence 'Jones is ill', the meta-linguistic sentence 'A is true' represents '"Jones is ill" is true', with 'A' in place of the quoted sentence 'Jones is ill'. As employed in making assertions within the meta-language or stating possibilities of truth or falsity, sentence schemata are thus to be invariably understood as standing in place of the names of the arbitrary sentences they represent.

So far we have referred only to particular sentences as expressing tautologies, contingent propositions, or contradictions. In a transferred sense we can also refer to sentence schemata representing the logical form of arbitrary sentences as expressing types of propositions. Some of the more important tautologies in this latter sense have been given by the logical tradition special names as "logical laws." Some of these are now listed:

i) $A \supset A$ law of identity
ii) $A \vee -A$ law of excluded middle
iii) $-(A \wedge -A)$ law of non-contradiction
iv) $--A \equiv A$ double negation law
v) $A \supset B \equiv -B \supset -A$ law of contraposition
vi) $-(A \wedge B) \equiv -A \vee -B$ DeMorgan's laws
vii) $-(A \vee B) \equiv -A \wedge -B$ DeMorgan's laws
viii) $A \supset (B \supset C) \equiv A \wedge B \supset C$ law of exportation

Laws such as iv)-viii) that are tautologous material equivalences are known as <u>logical equivalences</u>. The relation of logical equivalence between two propositions we represent by ' $\Leftrightarrow$ '. The fact that v) is a tautology may be thus stated as $A \supset B \Leftrightarrow -B \supset -A$. As was the case for entailment and material implication, logical equivalence as a relation between propositions must be carefully distinguished from the material equivalence sign as a logical connective.

<u>Exercises</u>

I. Determine by means of truth tables whether sentences of the following forms express tautologies, contingent propositions, or contradictions.

1. $(A \supset A) \supset A$
2. $A \supset -A$
3. $A \supset (B \supset A)$
4. $-(A \supset B \vee A)$
5. $(A \supset B) \vee (B \supset A)$
6. $(A \equiv B) \supset A \wedge -B$
7. $-(A \wedge B) \supset -(-A \vee -B)$
8. $(A \supset B) \supset (A \vee C \supset B \vee C)$
9. $A \vee B \supset (B \vee C \supset A \vee C)$
10. $(A \supset B) \wedge (C \supset D) \supset (A \wedge C \supset B \wedge D)$

II. Determine whether or not the following schemata express logical equivalences.

1. $(A \supset C) \vee (B \supset C) \equiv A \wedge B \supset C$
2. $A \supset (B \supset C) \equiv B \wedge -(A \supset C)$
3. $(A \equiv B) \equiv (-A \equiv -B)$
4. $A \vee (B \wedge C) \equiv A \wedge B \vee A \wedge C$
5. $A \vee (B \wedge C) \equiv A \vee B \wedge A \vee C$
6. $A \vee -A \equiv B \supset B$
7. $A \wedge (B \supset B) \equiv A$
8. $-(A \equiv B) \equiv (A \equiv -B)$

10. Disjunctive Normal Forms*

An alternate means for establishing whether a given sentence expresses a tautology can be developed by making use of a few logical laws established by the truth table procedure of the preceding section. In stating the laws we shall adopt two abbreviations that prove very helpful: the conjunction of two schemata $A \wedge B$ we shall write as the juxtaposition of A and B, or AB, and the negation of a schema we shall indicate by placing the negation sign over the schema, that is, writing $\bar{A}$ for $-A$. A contradiction will be symbolized by 'ϕ', a tautology by '$\square$'. In most cases the laws have two forms, one for conjunction and the other for disjunction. The basic laws are stated in the form of logical equivalences.

1. a) $AA \Leftrightarrow A$ b) $A \vee A \Leftrightarrow A$ idempotent laws
2. a) $AB \Leftrightarrow BA$ b) $A \vee B \Leftrightarrow B \vee A$ commutative laws
3. a) $A(BC) \Leftrightarrow (AB)C$ b) $A \vee (B \vee C) \Leftrightarrow (A \vee B) \vee C$ associative laws
4. a) $A(B \vee C) \Leftrightarrow AB \vee AC$ b) $A \vee BC \Leftrightarrow (A \vee B)(A \vee C)$ distributive laws
5. a) $\overline{(AB)} \Leftrightarrow \bar{A} \vee \bar{B}$ b) $\overline{(A \vee B)} \Leftrightarrow \bar{A}\bar{B}$ DeMorgan's laws
6. $\bar{\bar{A}} \Leftrightarrow A$ double negation law
7. a) $A\bar{A} \Leftrightarrow \phi$ b) $A \vee \bar{A} \Leftrightarrow \square$ negation laws
8. a) $A\square \Leftrightarrow A$ b) $A \vee \square \Leftrightarrow \square$ equivalence laws

The reader may recognize a correspondence between these laws and the axioms of Boolean algebra.[1]

The <u>disjunctive normal form</u> of a molecular schema with n atomic constituents is a disjunction of conjunctions, each conjunction containing each of the n atomic schemata or their negations. No conjunction can occur more than once within a disjunction. Any sentence schema can be transformed into its disjunctive normal form by means of the logical equivalences 1-8 just listed plus the following

> **Rule of Replacement:** If two schemata are logically equivalent, then they may be replaced for one another in any expression in which they may occur and still retain an expression logically equivalent to the original.

---

*This section may be omitted on a first reading of this chapter. It is essential only for an understanding of sections 22-24, 33, and 39.

As an application of the rule of replacement, consider the expression $A \supset B \vee C$. Since by the commutative law $B \vee C$ is equivalent to $C \vee B$, it can replace $B \vee C$ in $A \supset B \vee C$ to form the expression $A \supset C \vee B$. By the rule of replacement this derived expression is equivalent to the first.

The following steps provide an example of the expansion of the schema $A \vee B$ into its disjunctive normal form. The logical law used to derive each step from its predecessor by the rule of replacement is indicated to the right.

i) $A \vee B$
ii) $A\square \vee B\square$ equivalence law, 8a
iii) $A(B \vee \overline{B}) \vee B(A \vee \overline{A})$ negation law, 7b
iv) $(AB \vee A\overline{B}) \vee (BA \vee B\overline{A})$ distributive law, 4a
v) $AB \vee A\overline{B} \vee BA \vee B\overline{A}$ associative law, 3b
vi) $AB \vee A\overline{B} \vee AB \vee \overline{A}B$ commutative law, 2a
vii) $AB \vee AB \vee A\overline{B} \vee \overline{A}B$ commutative law, 2b
viii) $AB \vee A\overline{B} \vee \overline{A}B$ idempotent law, 1b

Since in the final step we have a disjunction of conjunctions, each containing A and B or their negations, we have the disjunctive normal form of $A \vee B$. Successive equivalences are derived by means of the rule of replacement. Step ii) follows from i) by virtue of the equivalence law that the conjunction of any schema with a tautology is equivalent to the schema itself. By the negation law a tautology is equivalent to the disjunction of a schema and its negation, and hence $B \vee \overline{B}$ and $A \vee \overline{A}$ can replace the tautology $\square$ in the right side of ii) to produce step iii). Similar justifications can be given for the remaining steps. The associative laws enable us to omit parentheses around conjunctions and disjunctions, since $A(BC) \Leftrightarrow (AB)C \Leftrightarrow ABC$ and $A \vee (B \vee C) \Leftrightarrow (A \vee B) \vee C \Leftrightarrow A \vee B \vee C$. Use of the associative law for disjunction is made in step v) in dropping the parentheses in iv).

As another example we expand the schema $\overline{(AB)}$ into its disjunctive normal form.

i) $\overline{(AB)}$
ii) $\overline{A} \vee \overline{B}$ DeMorgan's law, 5a
iii) $\overline{A}(B \vee \overline{B}) \vee \overline{B}(A \vee \overline{A})$ equivalence and negation laws
iv) $\overline{A}B \vee \overline{A}\overline{B} \vee A\overline{B} \vee \overline{A}\overline{B}$ distributive, associative, and commutative laws
v) $\overline{A}B \vee A\overline{B} \vee \overline{A}\overline{B}$ commutative and idempotent laws

Here we apply one of DeMorgan's laws in step ii), and then proceed as in the proceding expansion. This time steps are combined in order to abbreviate the expansion.

From the two expansions just presented a general procedure can be derived for expanding any schema with n atomic constituents into its disjunctive normal form. The steps to be followed are

1. If the schema is of the form of a negation of a conjunction or disjunction of schemata, then distribute the negation over the conjuncts (disjuncts) within parentheses and change conjunction (disjunction) to disjunction (conjunction) by DeMorgan's laws. This is done in step ii) of the preceding expansion where we infer from $\overline{(AB)}$ to $\overline{A} \vee \overline{B}$.

2. If the resulting schema is a disjunction of conjunctions that lack one of the n atomic schemata or its negation, then form conjunctions of the deficient conjunction with the disjunction of the missing schema and its negation, as in step iii) above where we infer from $\overline{A} \vee \overline{B}$ to $\overline{A}(B \vee \overline{B}) \vee \overline{B}(A \vee \overline{A})$.

3. Apply the distributive law to the resulting schemata and arrange the constituent conjuncts of the conjunctions in a consistent order. The result is a disjunction of conjunctions of the sort found in step iv) above.

4. Eliminate reiterated conjunctions. The result is a disjunction of conjunctions containing each of the n atomic schemata or its negation, with no conjunction occurring more than once. This is the disjunctive normal form of the schema.

If a material implication or equivalence connective occurs in the schema to be expanded, then we must first replace these connectives by the negation, conjunction, or disjunction connectives by means of definitions. We can define an implication $A \supset B$ as $\overline{A} \vee B$ and an equivalence $A \equiv B$ as $(A \supset B) \wedge (B \supset A)$ and hence as $(\overline{A} \vee B) \wedge (\overline{B} \vee A)$. Once these definitional equivalents have replaced the implication and equivalence connectives, the steps of the expansion procedure can be followed.

If the resulting disjunctive normal form of a schema with n atomic constituents is a disjunction of 2^n conjunctions, it will be a tautology when interpreted by the truth table procedure. If there are less than 2^n conjunctions (and the disjunctive normal form is not a contradiction), the schema will express a contingent proposition. These facts enable us to use the expansion procedure in order to determine whether or not a given schema expresses a tautology. We simply expand the schema to its disjunctive normal form and count the number of conjunctions. For example, we showed above that the disjunctive normal form of $A \vee B$ is $AB \vee A\overline{B} \vee \overline{A}B$ and that of $\overline{(AB)}$ is $\overline{A}B \vee A\overline{B} \vee \overline{A}\overline{B}$. Since both of these disjunctive normal forms have 3 conjunctions, and this is less than $2^2=4$, we know that the two schemata express contingent propositions. Similarly, since the disjunctive normal form of $A \supset (B \supset A)$ would prove to be $AB \vee A\overline{B} \vee \overline{A}B \vee \overline{A}\overline{B}$ with $2^2=4$ conjunctions, it would be shown to express a tautology.[2]

There is an important correspondence between the disjunctive normal form of a sentence schema and the truth tables of the previous section which enables us to read off the normal form directly from the truth table. Consider, for example, the truth table for $A \vee B$,

A	B	$A \vee B$
T	T	T
T	F	T
F	T	T
F	F	F

This tells us that whatever sentence $A \vee B$ represents, it will express a true proposition under the condition that A and B are true, A is true and B false, or A is false and B true. Now to say of a sentence within a meta-language that it is true, for example, '"It is raining" is true', is to say the same as one would in asserting within the object language 'It is raining'. Similarly, to say a sentence is false, for example, '"It is raining" is false', is to convey the same information as would be conveyed by asserting the negation 'It is not raining'. For the metalinguistic sentences in which we predicate truth or falsity of a mentioned sentence as named by quotation marks we can thus substitute as equivalent the sentence itself or its negation. Given this equivalence, we can substitute all occurrences in the above truth table where 'T' occurs by the schema itself and where 'F' occurs the negation of the schema. The truth table can be understood, therefore, as asserting $A \vee B$ under the condition that we may assert $AB \vee A\overline{B} \vee \overline{A}B$, the disjunction of conjunctions being equivalent to the first three rows of the truth table. This disjunction is the disjunctive normal form of $A \vee B$.

In general, any truth table showing the truth or falsity of a molecular schema as a function of the truth or falsity of its atomic constituents can be used to read off the schema's normal form. For every interpretation in which the schema is true we write an n-termed conjunction in which occurs each of the n atomic constituents or its negation, depending on whether the constituent is true or false for that interpretation. The disjunction of these conjunctions states the alternatives under which the schema is true, and is its disjunctive normal form. If the schema expresses a tautology and is true under all its 2^n possible interpretations, there will be 2^n conjunctions in the disjunctive normal form. If the schema is a contradiction, the only conjunctions appearing will be contradictions.

Exercises. Expand the following schemata into their disjunctive normal forms. Determine from the expansions whether or not the schemata express tautologies, contingent propositions, or contradictions.

1. A ∧ B
2. A ⊃ B
3. A ≡ B
4. A ⊃ A ∨ B
5. A ≡ A ∧ A
6. -(A ⊃ A)
7. A ∧ (B ∨ C)
8. A ⊃ B ∧ C
9. A ∧ C ⊃ (B ⊃ B)
10. A ∨ B ⊃ (C ≡ D)

NOTES FOR CHAPTER II

Section 5.

1. For the history of the development of this form of logic from the Stoics in ancient times through Frege and Russell see the Kneales [Development] and Bochenski [History].

Section 6.

1. We are adopting here a variation of the standard symbolism of Whitehead and Russell [P.M.]. An alternative called the "Polish notation" has been devised by Lukasiewicz [Logic] and is also used. In the Polish notation $p,q,r,\ldots$ are sentence schemata (or "propositional variables," as they are called) and connectives are represented by capital letters written to the left of the schemata which they operate on or combine. Thus, $-A$ in our symbolism becomes Np and $A \wedge B$ becomes Kpq. For a detailed development of systems of logic using this notation see Prior [Logic]. See also Iseminger [Logic] for an introduction with this notation that develops a natural deduction calculus.

Section 7.

1. The disparity between the logician's definition of material implication and the uses of 'if...then' in ordinary language has been often discussed. For two of the best discussions see Strawson [Logic Theory], pp. 82-90 and Neidorf [Deductive Forms], pp. 65-81.

Section 9.

1. Various ways of abbreviating the truth table decision procedures outlined here have been proposed. Among the more important are the methods of resolution in Quine [Methods], pp. 22-23 and of truth trees in Jeffrey [Logic], Ch. 4.

2. The relation of entailment is sometimes termed the relation of "implication" or "formal implication." It is called "strict implication" by Lewis [Survey] and symbolized by '$\prec$'. That this relation holds between names of sentences and not sentences themselves is argued in Quine [Logic], pp. 27-33.

Section 10.

1. A specific interpretation of Boolean algebra known as the <u>algebra of sets</u> (or classes) can be derived from the logical <u>laws just listed</u> by: 1) changing sentence schemata

to names of sets, 2) changing the relation of logical equivalence to the relation of identity between sets, 3) changing the connectives of negation, conjunction, and disjunction to the operations of complementation, intersection, and union on sets, and 4) changing the contradictory proposition to the null set and the tautologous proposition to the universal set. See Birkhoff and MacLane [Survey], pp. 336-338 or Arnold [Algebra] for introductory accounts of this algebra and its relationship to logic.

2. It can also be shown that every sentence schema can be expanded into a <u>conjunctive normal form</u>, a conjunction of disjunctions, each disjunction consisting of each of the constituent atomic schemata or its negation. The conjunction normal form of $A \wedge B$ would be $(A \vee B)(A \vee \bar{B})(\bar{A} \vee B)$. The expansion proceeds by replacing both A and B by the logically equivalent expressions $A \vee \phi$ and $B \vee \phi$ to form $(A \vee \phi)(B \vee \phi)$. ϕ is then replaced by the conjunction of the missing disjunct with its negation to produce $[A \vee (B\bar{B})][B \vee (A\bar{A})]$. The distributive law for disjunction is then applied and reiterated conjuncts dropped by the idempotent law for conjunction to produce the conjunctive normal form.

III. THE SENTENCE CALCULUS

11. Rules of Inference

When someone infers a conclusion from given premisses, we may regard him as acting in accordance with or following a <u>rule of inference</u>. For example, when the inference

1) If the war is being won, there is a decrease in enemy activity.
   If there is a decrease in enemy activity, the troops can come home.
   ∴ If the war is being won, the troops can come home.

is formulated, whoever presents it would be following a rule permitting him to infer from conditionals of the form $A \supset B$ and $B \supset C$ a conditional of the form $A \supset C$. This rule is general in the sense that it can be applied to an indefinite variety of inferences, only one of which is inference 1). It may, of course, be possible for someone to follow a rule such as that used for 1) without being able to explicitly formulate to himself or others the rule that he is following. To follow a rule requires only that we be able to reason ourselves in accordance with it and recognize successful applications and violations of the rule on the part of others. A person can be a skillful reasoner, correctly applying logical rules and detecting lapses in others, without being a student of logic.

These two features of rules of inference, their generality and their application by those who cannot necessarily explicitly formulate them, hold also for rules governing other types of activity. The legal rules (laws, local ordinances, etc.), moral rules, and customs governing our conduct are general and applied in a variety of different particular situations. General also are the rules of a game like chess (e.g., the rules for moving the pawn) or a sport like football (e.g., the rule prohibiting use of hands on offense). And like rules of inference we can follow or obey these rules without being able to explicitly formulate them to ourselves. The driver can obey the law requiring signalling before a turn without being able to explicitly state the law he is obeying, the chess player move his pieces correctly without being able to state to others the rules of chess.[1]

Why are there certain rules rather than others for a range of activity? Sometimes no specific reason can be given, it being quite arbitrary that one rule be favored over an alternative. The rule requiring drivers in the United States to drive on the right-hand side of the road would seem to offer an example of this. Some rule must be adopted in order to regulate traffic, but there seems no reason for preferring one requiring driving on the right to another requiring driving on the left. That the rule lacks a justification in terms of the purposes of driving is indicated by the adoption of driving on the left in other countries.

But it is evident that the rules governing our reasoning are not arbitrary and that a justification can be given for every rule of inference. To reason successfully we must be guaranteed that if our premises are true so is the conclusion drawn from them. This requires formulating inferences only according to rules that have this guarantee. We admit, therefore, as rules of inference only those for which it is impossible that inferences conducted in accordance with them have true premises and a false conclusion. Our truth table procedure of the previous chapter would show us that the rule governing inference 1) (called the rule of "hypothetical syllogism") satisfies this condition. Hence our acceptance of it as a justified rule. And in general the procedure by which we determine whether or not an inference is valid determines whether the inference is in accordance with a justified rule of inference. Rules of inference are thus in no sense arbitrary, as indicated by the fact that they do not differ from one nation or culture to another.

Another major point of difference between logical rules and those governing other activities is the interdependence relations holding for the former. Adopting certain rules of a game does not usually commit us to adopting any other specific rule. The rule requiring our saying 'Check' when checking the king, for example, in no way commits us to the rule permitting the king to move one space in all directions. They are in this sense independent, as are many other of the rules of the game. In contrast, adopting a rule of inference does commit us to others, and it is possible for us to show in a very precise way what these relations of interdependence are.

We proceed in the rest of this chapter to develop what is called a natural deduction calculus in which these relations of interdependence are determined. This calculus applied to inferences of the form evaluated in sentence logic is the sentence calculus. The interdependence between rules of inference is shown by means of proofs of rules from a restricted number of rules called "primitive rules." The proofs showing this dependence of some rules on others enables us to simplify and order the indefinite variety of rules used in deductive reasoning.[2]

It also enables us to introduce a significantly different aspect of the study of sentence logic. In the previous chapter we regarded the constituent sentences of an inference as expressing true or false propositions, and in terms of possible interpretations of these constituents decided on the validity or invalidity of given inferences. But in following a rule of inference we can disregard the interpretation of its constituents and apply the rule only to the form of expression of the inference. We apply the rule of hypothetical syllogism to inference 1), for example, only by virtue of the fact that the two premisses are conditionals of a certain form and that the conclusion is of the form of a conditional whose antecedent is the antecedent of the first premiss and whose consequent is the consequent of the second. Irrelevant to the application of the rule are the propositions expressed by these sentences and their truth or falsity. The justification of the rules, on the other hand, does require us to interpret the expressions to which the rule is being applied by means of the truth table procedure of the previous chapter. As we shall see, each of the rules introduced in the following sections can be justified in this manner. The general relation between rules of inference and their interpretation (between what is called the "syntax" and the "semantics" of logic) will be deferred until Section 30.

Finally, developing the sentence calculus is of importance in providing us with what is in many cases a simpler method of establishing the validity of an inference or the logical truth of a sentence. When the number of constituent sentences of an inference becomes very large, the truth table decision procedure becomes cumbersome and time-consuming. For example, an inference of the form $A \supset B$, $B \supset C$, $C \supset D$ ∴ $A \supset D$ would require a truth table with $2^4=16$ rows in order to be shown valid. As we shall see, the demonstration of its validity within the calculus is relatively simple.

12. Primitive Rules Governing Implication

We state now primitive rules governing expressions of the form of implications and present a method by which these rules can be applied within proofs. The rules stated permit us to introduce and eliminate the (material) implication connective. In subsequent sections introduction and elimination rules for negation and the other logical connectives will be stated. For heuristic reasons we present first the second rule of the calculus, the rule of implication elimination or (as it is more commonly known) modus ponens.

R2. Modus Ponens: From an implication and its antecedent we may infer the implication's consequent.

With A⊃B representing any implication, the rule may be formulated as

    1)   A⊃B
           A
           ―――
           B

In this form the rule could be used to permit inferences within a given object language. For example, the inference

    If the patient is well, his temperature is normal.
    The patient is well.
    ―――――――――――――――――――
    ∴ His temperature is normal.

could be licensed by the rule. Given the premisses of the inference the rule of <u>modus ponens</u> allows us to infer the conclusion.

By a <u>formal proof</u> is meant a sequence of schemata each of which is either assumed as a premiss or is derived as a consequence from previous schemata in the sequence by means of a rule of inference of the calculus. The terminal step of the sequence is the <u>conclusion</u> of the proof. The sequence of steps within a <u>proof</u> we number to the left of a proof. The line of a proof we indicate by means of a vertical line, with premisses being separated from the consequences derived from them by means of a horizontal line. Thus,

    1 | A
    2 | B
       |―
    . | .
    . | .
    . | .
    n | C

indicates the outline of a proof whose premisses are A and B and which terminates at step n with conclusion C. A proof of the conclusion exists when each of the steps 1,2,...,n is either a premiss or has been derived from preceding steps by means of the rules of the calculus.[1]

An example of a proof using the rule of <u>modus ponens</u> in the derivation of its conclusion is

    2) 1 | A   p
        2 | A⊃(B⊃C)   p
       3 | B⊃C   1,2, m.p.

Steps 1 and 2 are our premisses, as is indicated by writing 'p' to the right and by the horizontal line below them. Step 3 is the conclusion that follows from 1 and 2 by means of the rule of <u>modus ponens</u>. The justification for the step is written to the right, using an abbreviation for the rule. Note, however, that the rule of <u>modus ponens</u> formulated as 1) cannot be used to license the derivation of step 3 from 1 and 2. In 1) the rule is expressed in sentence schemata, whereas in the above proof the rule is <u>applied</u> to schemata. Our object language has become the symbols that had previously been part of our meta-language.

To apply rules such as <u>modus ponens</u> in proofs we must formulate them by means of <u>meta-schemata</u> $\phi, \psi, \chi, \ldots$. These are symbols that represent <u>arbitrary</u> (atomic or molecular) sentence schemata, just as sentence schemata represent arbitrary sentences. Using them <u>modus ponens</u> becomes reformulated as

$$3) \quad \begin{array}{c} \phi \\ \phi \supset \psi \\ \hline \psi \end{array}$$

with the horizontal line separating off the expressions to which the rule is applied (not necessarily premisses) from the expression that may be inferred by the rule. Step 3 in 2) is licensed in accordance with 3) when we take ϕ as standing for 'A' and ψ for 'B $\supset$ C'. In applying this rule it is understood that other schemata may occur between the antecedent ϕ and the implication $\phi \supset \psi$ to which the rule is applied. Also, there is no necessity that the schema ϕ occur first and $\phi \supset \psi$ second in the sequence of steps in the proof.

It should be obvious that the rule of <u>modus ponens</u> is a valid rule of inference, one which cannot ever allow us to derive a false consequence from previous steps that we interpret as true. That it is a valid rule can be verified by means of the following truth table:

ϕ	ψ	ϕ	$\phi \supset \psi$	ψ
T	T	T	T	T
T	F	T	F	F
F	T	F	T	T
F	F	F	T	F

For all possible interpretations of the sentence schemata that ϕ and ψ are taken to represent there will be none in which ϕ and $\phi \supset \psi$ are true and ψ false.

We are now in a position to state the first rule of the calculus, that permitting us to introduce the implication sign.

R1. __Implication Introduction__: _From a proof that a conclusion ψ may be derived from a premiss φ we may infer an implication of the form_ φ⊃ψ.

The assumption in R1 is a proof and not a premiss (or set of premisses), as in R2. To formulate and apply R1 we must introduce assumed proofs from which conclusions are inferred. These we call __subordinate proofs__ and indicate them by means of vertical lines to the right of the __main proof__, the proof in which occurs the conclusion being derived. Within a given proof there may be several subordinate proofs, or "sub proofs" as we shall refer to them. In this case there will be sub proofs relative to other sub proofs. For example, in a proof of the form

```
1  |A
2  |  |B
3  |  |  |C
.  |  |  |.
.  |  |  |.
.  |  |  |.
n-2|  |  |D
n-1|  |E
n  |F
```

the column headed by A is the main proof, since the final conclusion F occurs in it. Relative to this main proof is the sub proof headed by premiss B and terminating in E at step n-1. Relative to the first sub proof is a second sub proof with premiss C and conclusion D. There is no limit to the number of sub proofs that may occur within a given proof.

The proof assumed in stating the rule of implication introduction is a sub proof from which the implication is inferred with the proof to which it is relative. The symbolic formation of the rule becomes

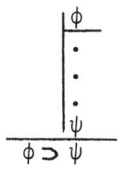

```
|φ
|.
|.
|.
|ψ
―――
φ⊃ψ
```

The vertical line indicates the assumed sub proof, with premiss φ and conclusion ψ. The main horizontal line separates this sub proof from the consequence φ⊃ψ within the proof to which the sub proof is relative. Given the sub proof in which ψ is derived from φ the rule of implication introduction permits us to infer the implication φ⊃ψ.[2] Taking φ to stand for 'A∧B' and ψ to stand for 'A∨B', we could use this rule to justify the derivation of the conclusion of the following proof:

```
1     │ A ∧ B      p
.     │   .
.     │   .
.     │   .
n-1   │ A ∨ B
n     │ A ∧ B ⊃ A ∨ B   1-n-1, ⊃ intro
```

Here we assume we have derived within a sub proof A ∨ B from A ∧ B by rules of the calculus in steps 1 through n-1. The conclusion in step n is inferred from this sub proof by the rule of implication introduction (abbreviated as "⊃ intro").

In order to apply the two rules governing the implication sign we need a means of using schemata that occur at an early stage in the proof (usually as a premiss) in a later stage. This is provided by adopting the

<u>Principle of Reiteration</u>: <u>Any step may be reiterated as a later step in the proof in which it occurs or into a sub proof relative to this proof.</u>

We usually use this principle in inferences of the form

```
1    │ A           p
2    │   │ B       p
.    │   │ .
.    │   │ .
.    │   │ .
i    │   │ A       1,r
.    │   │ .
.    │   │ .
n-1  │   │ C
n    │ D
```

where A is reiterated into a sub proof at step i and used in the proof of its conclusion C. Step i follows from 1 by the principle of reiteration. We abbreviate the justification as "1,r." We cannot, of course, reiterate a schema from a sub proof into a proof to which it is relative, for then any conclusion could be derived simply by assuming it. Reiteration is justified by the principle only within a given proof or into a proof subordinate to it.

The principle of reiteration should not be confused with the rules of inference of the calculus, being more basic than they. The justification for this principle lies in our being able to assume that if a sentence expresses a true proposition at one occurrence it will express a true proposition at every other occurrence.

To illustrate the use of the rules governing the implication sign and the principle of reiteration, we offer the following proof that A⊃C can be derived from A⊃B and B⊃C.

```
4)  1 | A ⊃ B         p
    2 | B ⊃ C         p
    3 |  | A          p
    4 |  | A ⊃ B      1,r
    5 |  | B          3,4,m.p.
    6 |  | B ⊃ C      2,r
    7 |  | C          5,6,m.p.
    8 | A ⊃ C         3-7, ⊃ intro
```

Note the use of a sub proof in steps 3-7. The existence of this sub proof warrants our inferring the conclusion by the rule of implication introduction. The proof that C follows from A in the sub proof is made possible by reiterating step 1 into the sub proof as step 4 and then inferring B by <u>modus ponens</u>. Our second premiss B⊃C is then reiterated in, allowing us to infer the conclusion of the sub proof C, again by <u>modus ponens</u>.

In such a proof we may adopt a strategy that greatly facilitates its construction. The conclusion of the proof has as its main connective an implication sign. We know that we can infer any implication provided we produce a proof that the consequent of the implication can be derived from its antecedent, the inference being then justified by the rule of implication introduction. Accordingly, in constructing proof 4) we can first set up the following outline of a proof:

```
 1  | A ⊃ B         p
 2  | B ⊃ C         p
 3  |  | A          p
    |  | .
    |  | .
    |  | .
n-1 |  | C
 n  | A ⊃ C         3-n-1, ⊃ intro
```

and then proceed to fill in steps 3-n-1 in the manner indicated in 4). Our strategy then is the following

Implication Strategy: If the main connective of a conclusion is an implication sign, then set up a sub proof whose premiss is the antecedent of the implication and whose conclusion is its consequent and infer the implication by the rule of implication introduction.

As for most of the proof strategies we shall introduce, this one is applied by inspecting the form of a conclusion and then constructing a sub proof from which the conclusion may be inferred.

As indicated above, there may be more than one sub proof within a given proof, and this may require more than one application of the Implication Strategy. For example, in the proof,

```
5)  1  |A⊃(B⊃C)           p
    2  | |A⊃B             p
    3  | | |A             p
    4  | | |A⊃(B⊃C)       1,r
    5  | | |B⊃C           3,4, m.p.
    6  | | |A⊃B           2,r
    7  | | |B             3,6, m.p.
    8  | | |C             7,5, m.p.
    9  | |A⊃C             3-8, ⊃ intro
   10  |(A⊃B)⊃(A⊃C)       2-9, ⊃ intro
```

two sub proofs are required, the first to prove the conclusion of the main proof, the second to prove the conclusion A⊃C of the first sub proof. To construct the proof we would first apply the Implication Strategy to the conclusion and set up a sub proof leading from A⊃B to A⊃C. We would then apply the strategy again to A⊃C, setting up a sub proof with premiss A and conclusion C. To show that C can be derived from A we simply reiterate in the premisses of our main proof and first sub proof and apply the rule of <u>modens ponens</u>. This second sub proof enables us to derive A⊃C as the conclusion of the first sub proof in step 9 by the rule of implication introduction, and the completing of the first sub proof enables the derivation of the conclusion in the main proof by the same rule.

<u>Exercises</u>. Construct proofs for the following, using the two implication rules and the principle of reiteration.
(The turnstile '⊦' is used to separate assumed premisses ϕ from the conclusion ψ that is to be derived from them. An expression of the form ϕ ⊦ ψ is thus to be read as 'From ϕ we can derive ψ.' The first exercise is to prove that from B as premiss we can derive A⊃B as conclusion. The fourth exercise is to prove that from the two premisses A⊃(B⊃C) and B we can derive A⊃C.)

1. B ⊦ A⊃B
2. A ⊦ B⊃B
3. A⊃(B⊃C) ⊦ B⊃(A⊃C)
4. A⊃(B⊃C),B ⊦ A⊃C
5. A⊃B ⊦ (B⊃C)⊃(A⊃C)

6. A⊃B ⊦ A⊃(C⊃B)
7. A ⊦ [A⊃(A⊃B)]⊃B
8. B⊃A,C⊃D ⊦ [A⊃(B⊃C)]⊃(B⊃D)
9. A⊃B ⊦ (C⊃D)⊃[(B⊃C)⊃(A⊃D)]
10. [(A⊃B)⊃C]⊃D ⊦ A⊃[B⊃(C⊃D)]

13. Rules for Negation, Conjunction, and Disjunction

We now state rules governing the use of the negation, conjunction, and disjunction connectives.

R3. Negation Introduction: From a proof that a contradiction may be derived from a premiss ϕ we may infer the negation of ϕ.

The assumption here, as for the rule of implication introduction, is that of a proof. The symbolic formulation of the rule thus requires a sub proof.

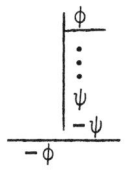

$$\frac{\begin{array}{|l} \phi \\ \vdots \\ \psi \\ -\psi \end{array}}{-\phi}$$

Here ψ and $-\psi$ constitute the contradiction we assume we have derived within a sub proof from our premiss ϕ. Assuming the derivation of this contradiction, the rule permits us to infer $-\phi$ within the proof to which the sub proof is relative. This rule is also known as the rule of "reductio ad absurdum" or the rule of "indirect proof." It is a rule frequently used in formal reasoning. To prove that a certain sentence is false we very often attempt to show that it leads to a contradiction.

The rule of negation introduction could be used to derive the conclusion in the following outline of a proof:

```
1   |-B         p
2   |    |A     p
.   |    |.
.   |    |.
.   |    |.
n-2 |    |B
n-1 |    |-B    1,r
n   |-A         2-n-1, - intro
```

If we could infer B in step n-2 from the premiss A of the sub proof, then after reiterating -B into the sub proof we would have shown that the contradictories B and -B can be derived from A. The rule of negation introduction then permits us to infer -A as the conclusion of the main proof.

R4. Double Negation Elimination: From the double negation of an expression ϕ we may infer ϕ itself.

$$\frac{--\phi}{\phi}$$

This rule simply states that negation applied to itself is self-cancelling. That this is a valid rule, that it is impossible for a schema of the form $--\phi$ to be true and ϕ false, can be easily verified by means of a truth table.

Rules R3 and R4 allow us to conduct proofs involving expressions in which the negation sign occurs. With them we can prove, for example, that the double negation of A can be derived from A. The proof utilizes the rule of negation introduction (abbreviated as "- intro").

```
1)  1 |A          p
    2 | |-A       p
    3 | | |A      1,r
    4 | | |-A     2,r
    5 | --A       2-4, - intro
```

Since the premiss -A of the sub proof has been shown to lead to a contradiction, we may infer --A by negation introduction. As for the proofs employing the rule of implication introduction, this is constructed by noting the form of the conclusion and then constructing a sub proof from which the conclusion can be derived. We know that by negation introduction --A can be inferred if -A can be shown to lead to a contradiction. Accordingly, we assume -A in a sub proof and attempt to derive the contradiction.

By means of both negation rules R3 and R4 we can also prove that B can be derived from A and its negation -A, or that any sentence can be derived from a contradiction.

```
2)  1 |A          p
    2 |-A         p
    3 | |-B       p
    4 | | |A      1,r
    5 | | |-A     2,r
    6 | --B       3-5, - intro
    7 |B          6, -- elim
```

Our strategy here is to assume -B as a premiss in order to show that it leads to a contradiction. This allows us to infer --B by negation introduction in step 6, and then our conclusion by double negation elimination (abbreviated as "-- elim") in step 7. Again our procedure is to note the form of the conclusion and construct a sub proof from which it may be derived by the rules made available to us.

As seen in the construction of proofs 1) and 2) there are two different strategies for use of the negation rules, one used when the conclusion is of the form of a negation, the other when it is positive. These are combined in the following

Negation Strategy: a) If the conclusion is of the form of a negation, then set up a sub proof showing the expression that is negated leads to a contradiction. b) If the conclusion is positive, it may be convenient to set up a sub proof showing that its negation leads to a contradiction. If it does, the negation of this negation can be inferred by negation introduction, and then the conclusion itself by double negation elimination.

Note that in b) we only say "it may be convenient." This is a low priority strategy that should be resorted to only if none other is applicable. For example, if our conclusion were of the form of an implication, the Implication Strategy would have priority over Negation Strategy b).

Our negation rules can be combined with the implication rules to conduct proofs involving expressions in which both implication and negation connectives occur. Such would be the following:

```
3)  1  | A ⊃ B          p
    2  |  | -B           p
    3  |  |  | A         p
    4  |  |  | A ⊃ B     1,r
    5  |  |  | B         3,4, m.p.
    6  |  |  | -B        2,r
    7  |  | -A           2-6, - intro
    8  | -B ⊃ -A         2-7, ⊃ intro
```

This proof is accomplished by combining our Implication and Negation Strategies. The Implication Strategy leads from an inspection of the conclusion in the main proof to the setting up of the first sub proof with premiss -B and conclusion -A in order to infer the conclusion by implication introduction. Since the conclusion of this sub proof is a negated expression, our Negation Strategy a) dictates the second sub proof in which we (successfully) attempt to show that a contradiction can be derived from A.

We now state four rules for introducing and eliminating the conjunction and disjunctive connectives.

R5. <u>Conjunction Introduction</u>: <u>From a schema ϕ and a schema ψ, we may infer their conjunction.</u>

$$\frac{\phi}{\psi}$$
$$\overline{\phi \wedge \psi}$$

R6. **Conjunction Elimination**: From a conjunction $\phi \wedge \psi$ we may infer either of its conjuncts ϕ or ψ.

$$\frac{\phi \wedge \psi}{\phi}$$
$$\psi$$

This rule actually combines two rules: that ϕ may be inferred from $\phi \wedge \psi$, and that ψ may also be inferred from the conjunction. Writing two conclusions under the horizontal line is a convenient way of summarizing the two rules.

R7. **Disjunction Introduction**: From a schema ϕ we may infer the disjunction of ϕ with any other schema ψ.

$$\frac{\phi}{\phi \vee \psi}$$
$$\psi \vee \phi$$

Again we have a combination of two rules, for ϕ can occur as either the first or second disjunct of the inferred expression.

The validity of rules R5-R7 can again be easily verified by the reader. It is obvious that the rules directly reflect the truth table definitions of conjunction and disjunction given in Section 6.

R8. **Disjunction Elimination**: From a disjunction $\phi \vee \psi$ and proofs that a third schema χ may be derived from both disjuncts, we may infer χ.

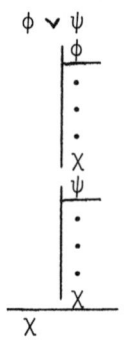

$$\phi \vee \psi$$
$$\begin{array}{|l} \phi \\ . \\ . \\ . \\ \chi \end{array}$$
$$\begin{array}{|l} \psi \\ . \\ . \\ . \\ \chi \end{array}$$
$$\overline{\chi}$$

This rule licenses the derivation of the consequence χ on the assumption of a disjunction and two sub proofs whose conclusions are both χ. It is understood that these sub proofs are independent of one another, that steps in the first cannot be reiterated as steps in the second. The justification for this complicated rule would be that since one of the disjuncts of $\phi \vee \psi$ is assumed to be true (by definition of the disjunction sign), if a conclusion is derivable from both disjuncts it must be true also.

As an example of a proof using the conjunction and disjunction rules we offer the following:

```
4)   1  │A ∨ (B ∧ C)           p
     2  │ │A                    p
     3  │ │A ∨ B                2, ∨ intro
     4  │ │A ∨ C                2, ∨ intro
     5  │ │(A ∨ B) ∧ (A ∨ C)    3,4, ∧ intro
     6  │ │B ∧ C                p
     7  │ │B                    6, ∧ elim
     8  │ │A ∨ B                7, ∨ intro
     9  │ │C                    6, ∧ elim
    10  │ │A ∨ C                9, ∨ intro
    11  │ │(A ∨ B) ∧ (A ∨ C)    8,10, ∧ intro
    12  │(A ∨ B) ∧ (A ∨ C)      1,2-5,6-11, ∨ elim
```

We prove the conclusion from the premiss and the two sub proofs by the rule of disjunction elimination. The ϕ of our rule is taken here to stand for 'A', ψ for 'B ∧ C', and χ for '(A ∨ B) ∧ (A ∨ C)'. The conclusion (A ∨ B) ∧ (A ∨ C) can be derived by means of disjunction elimination in step 12 because in the two sub proofs it is derived from the disjuncts A and B ∧ C of the premiss of the main proof. The rules of conjunction introduction and elimination and disjunction introduction are used within the two sub proofs.

Invariably when there is a disjunction given, in order to infer a consequence from it we will need to employ the rule of disjunction elimination. Accordingly, we formulate a strategy for using the disjunction elimination rule.

   Disjunction Strategy: If the main connective of a
   step from which we want to derive a given conclusion
   is the disjunction sign, then employ the rule of
   disjunction elimination by setting up two sub proofs
   in which the premisses are the disjuncts and the
   conclusions the conclusion that is to be derived.

In applying this strategy to proof 4) we would first set up the following outline of a proof:

```
    1   │A ∨ (B ∧ C)           p
    2   │ │A                    p
    ⋮   │ │ ⋮
    i   │ │(A ∨ B) ∧ (A ∨ C)
    j   │ │B ∧ C                p
    ⋮   │ │ ⋮
   n-1  │ │(A ∨ B) ∧ (A ∨ C)
    n   │(A ∨ B) ∧ (A ∨ C)      1,2-i,j-n-1, ∨ elim
```

We would then proceed to fill in the intermediate steps 2-i and j-n-l within the sub proofs.

Note that the disjunction strategy is the only one of the strategies that is applied to a step at the "top" of a proof. In constructing a proof we usually proceed with regard to the conclusion that is to be inferred. Note also that each of the rules of inference formulated so far that assume a proof (R1, R3, and R8) has a strategy for its use. In general, a sub proof should never be constructed unless in accordance with a definite strategy.

The conjunction and disjunction rules can be combined with the implication and negation rules to conduct proofs. Examples of this are included in the exercises given below.

<u>Exercises</u>. Prove the following.

1. -A ⊢ A ⊃ B
2. A ⊃ B, -B ⊢ -A
3. -A ⊢ A ⊃ -A
4. A, B ⊢ -(A ⊃ -B)
5. -B ⊃ -A ⊢ A ⊃ B
6. A ∧ B ⊢ B ∧ A
7. A ⊃ (B ⊃ C) ⊢ A ∧ B ⊃ C
8. A ⊃ B ⊢ A ∧ C ⊃ B
9. A ⊃ B ∧ C ⊢ A ⊃ B
10. A ⊃ B, C ⊃ D ⊢ A ∧ C ⊃ B ∧ D

11. A ∨ B ⊢ B ∨ A
12. A ⊃ B ⊢ A ⊃ B ∨ C
13. A ∨ C ⊃ B ⊢ A ⊃ B
14. (A ⊃ B) ∨ (A ⊃ C) ⊢ A ⊃ B ∨ C
15. (A ⊃ C) ∨ (B ⊃ C) ⊢ A ∧ B ⊃ C
16. A ⊃ B, C ⊃ D, A ∨ C ⊢ B ∨ D
17. A ⊃ B, C ⊃ D, -B ∨ -D ⊢ -A ∨ -C
18. A ∨ B, -A ⊢ B
19. -(A ∨ B) ⊢ -A ∧ -B
20. -(A ⊃ B) ⊢ A ∧ -B

14. <u>Derived Rules of Inference and Applications</u>

<u>Derived Rules of Inference</u>. A proof that a conclusion can be derived from one or more assumed premises we call a <u>conditional proof</u>. The proofs carried out in the examples and exercises of the two preceding sections were all of this variety. Once such a proof has been completed it is possible to adopt its result as a rule of inference to be used to license steps in future proofs in the manner of our primitive rules. Every conditional proof in this way becomes the proof of a rule of inference licensing the inference from the proof's premises to its conclusion. Rules proved in this manner by means of the primitive rules we call <u>derived rules of inference</u>.

As an example, we proved in 2) of the previous section that B can be derived from A and -A. In order to apply this result to sentence schemata within a proof we change the schemata A and B into meta-schemata. The derived rule is now stated as

$$\frac{\phi \quad -\phi}{\psi}$$

We shall call this the "rule of negation elimination" (or "- elim"). With it we can simplify the proof of exercise 1 of the preceding section in the following manner:

```
1 |-A       p
2 | |A      p
3 | | |-A   1,r
4 | | |B    2,3, - elim
5 |A ⊃ B    2-4, ⊃ intro
```

The derived rule is used in step 4 to license the inference of B from A and -A. This enables us to dispense with a second sub proof previously required, and reduces the number of steps from 8 to 5. In doing this we make use of previous proofs, much as the mathematician uses theorems that have been proved in the proofs of further theorems. Notice that derived rules are not required for proofs. Their function is that of simply reducing the number of steps within a proof that without them would be necessary.

Some of the more commonly used rules of inference that have been derived either as examples or exercises in the preceding section are now listed. Abbreviations that may be helpful are indicated.

DR1. Hypothetical Syllogism (h.s.)

$$\frac{\phi \supset \psi \quad \psi \supset \chi}{\phi \supset \chi}$$

DR2. Double Negation Introduction (-- intro)

$$\frac{\phi}{--\phi}$$

DR3. Modus Tollens (m.t.)

$$\frac{\phi \supset \psi \quad -\psi}{-\phi}$$

DR4. Disjunctive Syllogism (d.s.)

$$\frac{\phi \vee \psi \quad -\phi}{\psi}$$

DR5. Negation Elimination
       (- elim)

$$\frac{\phi \quad -\phi}{\psi}$$

DR7. Negation Disjunction
       Elimination (-∨ elim)

$$\frac{-(\phi \vee \psi)}{-\phi \wedge -\psi}$$

DR6. Contraposition
       (contrapos)

$$\frac{\phi \supset \psi}{-\psi \supset -\phi}$$

DR8. Negation Implication
       Elimination (-⊃ elim)

$$\frac{-(\phi \supset \psi)}{\phi \wedge -\psi}$$

What we adopt as a primitive rule and what a derived rule is, of course, to a great degree arbitrary. By adopting DR7, the rule of negation disjunction elimination, as primitive, for example, we could derive the rule of disjunction introduction. It is arbitrary, then, which of these rules we consider as primitive and which derived, the decision being made on the basis of simplicity. What the sentence calculus enables us to see is the interdependence between rules of inference, how adopting certain rules commits us to adopting others derivable from them. It also enables us to see how the denial of a certain rule of inference forces us to deny other rules or adopt them as primitive. In the intuitionistic calculus of Heyting, for example, the rule of double negation elimination, $--\phi \vdash \phi$, is denied on the basis that one who has proved that the negation of an expression ϕ leads to a contradiction has not proven ϕ itself.[1] But if we exclude double negation elimination from our primitive rules it will be impossible to derive (among others) DR5, negation elimination, and DR4, the disjunctive syllogism. If these rules are to be retained, the rule of negation elimination must then be taken as primitive. The rules of inference of the calculus thus form a connected system. Rejecting a rule or altering one of them will have an effect on other rules within this system.

Justification of the Rules. We mentioned in Section 11 that rules governing any type of activity can often be justified in terms of the purposes of this activity. In using a deductive inference we offer our guarantee that the conclusion is true if the premises are. A rule of inference will be justified and valid, therefore, if it is such that someone following the rule will never infer a false conclusion from true premises.

Most of the rules introduced can be easily shown to be valid by a direct application of the truth table decision procedure of the previous chapter. As we have seen, the rules of _modus ponens_, double negation elimination, conjunction introduction and elimination, and disjunction introduction can all be shown to be valid in this manner. The validity of these rules is shown by establishing an entailment relation between the propositions expressed by the schemata to which the rule is applied and the consequence derived from them. For example, the disjunction introduction rule $\phi \vdash \phi \vee \psi$ is valid, since we can establish that $\phi \Rightarrow \phi \vee \psi$.

The validity of the rule of implication introduction cannot, however, be justified in this manner. The rule assumes a proof that an expression ψ is provable from ϕ. In order to prove ψ each step in the derivation must be licensed by primitive rules which we can show to be valid. If these are valid, each step will be entailed by preceding steps. And since entailment is a transitive relation, that is, if $\phi \Rightarrow \chi$ and $\chi \Rightarrow \psi$ then $\phi \Rightarrow \psi$. the conclusion ψ will be entailed by ϕ. But if ϕ entails ψ, then it will be impossible for $\phi \supset \psi$ to be false by definition of the material implication sign and the entailment relation. Hence the rule is justified: it is impossible for $\phi \supset \psi$ to be false, assuming that ψ can be derived from ϕ. A justification that relies on this can be given for the rules of negation introduction and disjunction elimination. The rule of negation introduction assumes a sub proof leading from ϕ to the contradiction ψ and $-\psi$. By the implication introduction and conjunction introduction rules we can infer from this sub proof $\phi \supset \psi \wedge -\psi$. Our truth table procedure can now establish $\phi \supset \psi \wedge -\psi \Rightarrow -\phi$, and hence the validity of the rule. Similarly, by implication introduction we can infer from the sub proofs of the disjunction elimination rule $\phi \supset \chi$ and $\psi \supset \chi$. By the truth table procedure we can now establish that $(\phi \vee \psi) \wedge (\phi \supset \chi) \wedge (\psi \supset \chi) \Rightarrow \chi$.

Having established that the primitive rules are valid, it follows that any rule derivable from them is valid also. Every derived rule will be proved in a sequence of steps each either a premiss or licensed by a primitive rule or by the reiteration principle. Each step not a premiss is thus either entailed by preceding steps or (where the rule is applied to a sub proof) follows by definition of the material implication sign. It follows that the premisses of the proof entail the conclusion and that the derived rule is therefore valid.

<u>Evaluation of Inferences</u>. Having justified our rules of inference in this manner, it is an easy matter to see how the rules can be used to evaluate inferences formulated within a given object language. Any inference can be evaluated as valid if it is formulated in accordance with a valid rule. The task of evaluation thus becomes that of showing that an inference is in accordance with a primitive or derived rule of the calculus.

For an inference such as 1) of Section 11 this evaluation can be done directly. The inference is formulated in accordance with our previously proved DR1, the hypothetical syllogism, and is therefore valid. For an inference such as

> 1) If our team wins tonight, we will celebrate.
> <u>If our team doesn't win, we will stay at home.</u>
> ∴ Either we will celebrate or stay at home.

we must prove that the inference is in accordance with a derived rule. The rule governing 1) is clearly

$$\frac{A \supset B}{\frac{-A \supset C}{B \vee C}}$$

Since we can construct a conditional proof in which the conclusion $B \vee C$ is derived from the premises, we know that this can serve as a derived rule of the calculus (after changing sentence schemata to meta-schemata). We know therefore that inference 1) is valid. Whenever we have a conditional proof of a derived rule governing a given inference we have in this manner a proof of the inference's validity.

Note, however, that this method does not enable us to prove that an inference is invalid. In general, there is no way of proving that a certain conclusion cannot be derived from given premises. For if we cannot prove a conclusion this means one of two things: 1) either no proof is possible and the inference in question is invalid; or 2) we have not been clever enough to discover a proof which can in principle be discovered. Since we have no way of knowing in a given case which of these two alternatives hold, we have no right to judge the inference invalid in the absence of a proof. The calculus thus gives us only a method for deciding that an inference is valid. For a method that decides whether inferences are invalid as well as valid, or what is known as an <u>effective decision procedure</u>, we must resort to the more cumbersome truth table decision procedure of the previous chapter.

<u>Exercises</u>. Prove the validity of the following inferences by means of the sentence calculus. Use derived rules DR1-8 where helpful.
    1. Inferences 2, 3, 7, 8, 11, 12, 14, and 15 of Exercises II for Section 8.
    2. Logic is either exciting or dull. If it were dull, then few would study it. If few studied it, there would not be many logic texts. But there are many logic texts. Therefore, logic is exciting.
    3. There would not be many logic texts unless logic were exciting. But there are many logic texts. If logic is exciting then it is not dull. Hence, unless few don't study logic it is not dull.
    4. Logic is exciting only if it has applications. If logic is not exciting then it is dull. Therefore, if logic has no applications, then it is dull if few study it.

5. That many study logic is a necessary condition for its being exciting. If many study logic the subject is useful because of many applications for it. But there are no applications for logic. Logic is either exciting or many students have been duped. And so we must conclude that many students have been duped.

6. We should conclude that logic is not dull. For suppose that many students have been duped. It would follow that the subject is neither useful nor has applications. But the subject does have applications, and logic would not be dull unless students had been duped.

7. Clearly, logic is dull. For suppose that logic is not dull. In that case, people would be more interested in the words of Bertrand Russell than in those of Martha Mitchell and Playboy would be outsold by the Notre Dame Journal of Symbolic Logic. But, as everyone knows, Playboy outsells that scholarly journal.

8. Antigone reasoned soundly only if Natural Law is supreme. But unless Socrates erred in his conversation with Crito, Natural Law is not supreme. Of course, if Socrates erred, then if Antigone reasoned soundly, our boys can be withdrawn from Thailand and Vietnam. Thus, Antigone reasoned soundly only if our boys can be withdrawn from Vietnam.

9. We can conclude that if two gases have the same temperature and volume, they must have the same pressure. For suppose that two gases are at the same temperature. Then it would follow that their molecules have the same kinetic energy. Now we know that equal volumes of two gases contain the same number of molecules, and that the pressures of two gases are equal if the numbers of molecules and their kinetic energies are equal. Hence our conclusion.

10. If forces F_1 and F_2 are equal and are the only forces acting on point p, the resultant force F_x will either be zero or will have the same direction as the bisector of the angle between forces F_1 and F_2. If F_x is zero, then F_1 and F_2 are equal and directly opposed. It follows that if F_1 and F_2 are equal but not directly opposed, then they are not the only forces acting on p.

15. Categorical Proofs

In a conditional proof we prove a conclusion by deriving it by means of primitive or derived rules from given premises. We can also conduct proofs in which a conclusion is derived only from sub proofs, with no premisses occurring in the main proofs. Such proofs are called <u>categorical</u> <u>proofs</u>. As an example, it is possible to construct a categorical proof of $-(A \wedge -A)$.

```
1 |  A ∧ -A        p
2 |  A             1, ∧ elim
3 |  -A            1, ∧ elim
4  -(A ∧ -A)       1-3, - intro
```

Notice that there are no premisses in the main proof here. The conclusion is derived only from the sub proof showing that A∧-A leads to a contradiction. All categorical proofs are constructed by noting the form of the conclusion and using either the Implication, Negation, or Equivalence Strategies (the latter to be discussed shortly). If the conclusion is positive and not of the form of an implication or equivalence, Negation Strategy b) is invariably used.

Because of the justification that can be given to the primitive rules of the calculus every conclusion of a categorical proof expresses a tautology when interpreted. If the sub proofs from which the conclusion is derived have been constructed in accordance with the rules of the calculus, it is impossible that the conclusion is false. The calculus can therefore be employed as a means of establishing that a certain sentence expresses a tautology. First we represent the sentence by a sentence schema. Then we attempt to prove this schema by means of a categorical proof. If we are successful, we have established that the sentence expresses a tautology. But parallel to the situation holding for evaluated inferences, failure to construct a proof does not establish that the sentence does not express a tautology, that it is either contingent or contradictory. Our failure might be due to our inability to discover a proof that could in principle be constructed.

Rules governing (material) equivalence. We still lack introduction and elimination rules for the (material) equivalence connective.

R9. Equivalence Introduction: From a proof that ψ can be derived from a premiss ϕ and a proof that ϕ can be derived from ψ we may infer the equivalence $\phi \equiv \psi$.

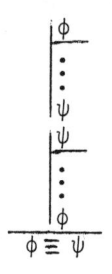

$$
\begin{array}{|l}
\phi \\
\vdots \\
\psi \\
\hline
\end{array}
\quad
\begin{array}{|l}
\psi \\
\vdots \\
\phi \\
\hline
\end{array}
$$
$$\overline{\phi \equiv \psi}$$

R10. Equivalence Elimination: From the equivalence $\phi \equiv \psi$ we may infer the implications $\phi \supset \psi$ and $\psi \supset \phi$.

$$\frac{\phi \equiv \psi}{\phi \supset \psi}$$
$$\psi \supset \phi$$

The rule of equivalence introduction is important for its application in categorical proofs of schemata that express logical equivalences. One such proof is offered now. We make use in it of our derived as well as primitive rules.

```
2)   1  | -(A∧B)                    p
     2  |  | -(-A∨ -B)              p
     3  |  | --A ∧ --B              2, - ∨ elim (DR7)
     4  |  | --A                    3, ∧ elim
     5  |  | --B                    3, ∧ elim
     6  |  | A                      4, -- elim
     7  |  | B                      5, -- elim
     8  |  | A∧B                    6,7, ∧ intro
     9  |  | -(A∧B)                 1,r
    10  | --(-A∨-B)                 2-9, - intro
    11  | -A∨ -B                    10, -- elim
    12  | -A∨ -B                    p
    13  |  | -A                     p
    14  |  |  | A∧B                 p
    15  |  |  | A                   14, ∧ elim
    16  |  |  | -A                  13,r
    17  |  | -(A∧B)                 14-16, - intro
    18  |  | -B                     p
    19  |  |  | A∧B                 p
    20  |  |  | B                   19, ∧ elim
    21  |  |  | -B                  18,r
    22  |  | -(A∧B)                 19-21, - intro
    23  | -(A∧B)                    12,13-17,18-22, ∨ elim
    24  -(A∧B) ≡ -A∨ -B             1-11,12-23, ≡ intro
```

Since our conclusion is an equivalence we begin this proof by setting up two sub proofs, one leading from -(A∧B) to -A∨ -B, the other from -A∨ -B to -(A∧B). This is an application of our final strategy for the sentence calculus.

> Equivalence Strategy: If the conclusion is of the form of an equivalence, then use the rule of equivalence introduction by setting up two sub proofs, one headed by the left side of the equivalence and terminating in the right side as its conclusion, the other headed by the right side and terminating in the left.

The proof 2) is constructed in three stages:
i) application of the Equivalence Strategy to set up the first two sub proofs; ii) application of the Negation Strategy b) on the top sub proof and the Disjunction Strategy on the bottom sub proof; and iii) application of Negation Strategy a) on the two sub proofs constructed for the bottom sub proof. Schematically these stages are

```
        |-(A∧B)              |-(A∧B)                    |-(A∧B)
        | |-(-A∨-B)          | |-(-A∨-B)                | |-(-A∨-B)
        | | .                | | .                      | | .
        | | .                | | .                      | | .
        | | .                | | .                      | | .
        | |--(-A∨-B)         | |--(-A∨-B)               | |--(-A∨-B)
        | -A∨-B              | -A∨-B                    | -A∨-B
        | -A∨-B              | -A∨-B                    | -A∨-B
        |                    | |-A                      | |-A
        |                    | | .                      | | .
 i)     | .           ii)    | | .            iii)      | |A∧B
        | .                  | | .                      | | .
        | .                  | |-(A∧B)                  | |-(A∧B)
        |                    | -B                       | -B
        |                    |  .                       | |A∧B
        |                    |  .                       | | .
        |                    |  .                       | | .
        |                    | |-(A∧B)                  | |-(A∧B)
        |-(A∧B)              |-(A∧B)                    |-(A∧B)
        -(A∧B)≡-A∨-B         -(A∧B)≡-A∨-B              -(A∧B)≡-A∨-B
```

The steps indicated by the dots in iii) are then filled in to complete the proof. Using the strategies in this manner enables us to resolve a complicated proof into a series of relatively simple proofs.

Exercises. Construct categorical proofs for the following. (The turnstile '⊢' to the left of a schema φ is to be interpreted to mean 'φ is derivable relative to no premisses.')

1. ⊢ A∨-A
2. ⊢ A⊃(B⊃A)
3. ⊢ -A⊃(A⊃B)
4. ⊢ (A∨C⊃B)⊃(A⊃B)
5. ⊢ A∨A≡A
6. ⊢ A⊃-A≡-A
7. ⊢ A⊃(B⊃C)≡A∧B⊃C
8. ⊢ A⊃B≡-(A∧-B)
9. ⊢ A⊃B≡-A∨B
10. ⊢ A∧(A∨B)≡A∨A∧B
11. ⊢ A∧(B∨C)≡A∧B∨A∧C
12. ⊢ A∨B∧C≡(A∨B)∧(A∨C)
13. ⊢ -(A∨B)≡-A∧-B
14. ⊢ A∧(B∨-B)≡A
15. ⊢ A∨B∧-B≡A
16. ⊢ A∨(B∨-B)≡B∨-B
17. ⊢ A∨-A∧B≡A∨B
18. ⊢ (A≡B)≡(-A≡-B)
19. ⊢ -(A≡B)≡(A≡-B)
20. ⊢ (A≡B)≡A∧B∨-A∧-B

NOTES FOR CHAPTER III

Section 11.

    1. For a general discussion of reasoning as a species of rule-governed activity see Ryle [Mind], Chs. II and IX.

    2. The natural deduction calculus was discovered independently by Gentzen and Jaskowski. For a discussion of this calculus and its importance in the development of logic see the Kneales [Development], pp. 538-548. The rules formulated here are almost exactly those originally adopted by Gentzen.

Section 12.

    1. The use of vertical and horizontal lines in order to apply Gentzen's rules is due to Fitch [Logic].

    2. It is understood that it may be necessary to reiterate steps from proofs to which the sub proof is relative into the proof in which ψ is derived from ϕ. For heuristic reasons we do not make this assumption explicit. It would be made explicit by adopting as our rule

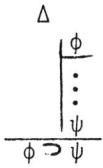

$$\begin{array}{c} \Delta \\ \phi \\ \vdots \\ \psi \\ \hline \phi \supset \psi \end{array}$$

where Δ is a sequence of steps in the proof to which the sub proof from ϕ to ψ is relative. The same assumption is made for rules R3, R8, and R9, the other primitive rules that follow employing sub proofs. For formulations of the rules making explicit this assumption as well as the assumption that the steps to which rules are applied need not be premises and may be separated by intermediate steps see Leblanc [Deductive Inference], Sec. 1.2.

Section 14.

    1. See Heyting [Intuitionism], pp. 97-105. In Fitch's calculus ([Logic], pp. 53ff) the rule of negation introduction is rejected, thus requiring negation elimination and double negation introduction to be taken as primitive.

IV. PREDICATE LOGIC: BASIC CONCEPTS

16. Terms

The Scope of Predicate Logic. So far we have been considering only inferences whose validity depends on recurrences of constituent sentences expressing the same proposition. But there is a wide variety of inferences whose evaluation requires consideration of recurrences of a totally different type of element. To illustrate the inadequacy of the methods of sentence logic consider the inference,

   1) All men are mortal.
     All Greeks are men.
     ∴ All Greeks are mortal.

As represented within sentence logic the form of 1) would be A,B ∴ C. Since this reveals no connection between premisses and conclusion the inference would be evaluated as invalid. But 1) is clearly valid. There must, then, be recurring elements within the inference other than sentences upon which its validity depends. And a glance at inference 1) reveals what these elements are. In the first and second premisses we find different tokens of the term 'men', in the first premiss and the conclusion tokens of 'mortal', and in the second premiss and conclusion the term 'Greeks' recurs. When we analyze the sentences of the premisses and conclusion into their constituent terms, we thus find elements which provide the required connection between premisses and conclusion.[1]

But as before for sentence logic we must immediately qualify ourselves and add that what is essential for the validity of an inference like 1) is not the recurrence of terms themselves. It is instead recurrences of terms sharing a common content, that is, terms expressing a common meaning or standing for the same objects for one who understands them. For example, in the inference,

   2) All men are mortal.
     Socrates is a human being.
     ∴ The most famous teacher of Plato will die.

there are no recurring terms as linguistic expressions. But since 'men' and 'human being' have the same meaning for us, as do 'mortal' and 'will die', there is a connection between premisses and conclusion. This is also insured by the nature of the terms 'Socrates' and 'the most famous teacher of Plato'. Because the name and the descriptive phrase are presumably understood by us to stand for the same individual, there is again a recurrence, though not of linguistic expressions.

We may define <u>predicate logic</u> as that branch of logic which evaluates inferences whose validity depends on a connection between terms having the same content. As before for sentence logic, the evaluation procedure for inferences can be extended to a general decision procedure evaluating the logical status of molecular sentences. Here the scope of predicate logic extends to all sentences whose logical status depends on the recurrence of constituent terms with the same content. Before taking up the various decision procedures for predicate logic in Chapter V, however, we must first consider the nature of the constituent terms of sentences and the logical symbolism used to represent sentences analyzed into terms.

<u>Singular</u> and <u>General</u> <u>Terms</u>. The primary classification of terms is on the basis of the quantity of objects to which they can be ascribed. Those expressing attributes that are true of but a single object, e.g. 'Socrates' and 'the Empire State Building', are called <u>singular terms</u>; those that can be ascribed to a plurality of individual objects, e.g. 'red' and 'table', are called <u>general</u> terms. A term is said to <u>denote</u> those objects to which it can be truly ascribed.[2] Thus, the singular term 'Socrates' denotes the historical individual who is the central figure of the Platonic dialogues, while 'red' denotes the class of all those objects that are red. The object or objects denoted by a term is said to constitute the term's <u>extension</u>. This extension has been traditionally contrasted to the term's <u>intension</u>, the attributes or relations used in identifying objects to which the term is ascribed. The intension of a singular term is thus the attributes we would use in identifying the single object denoted by the term. The intension of 'Socrates', for example, might be the attributes of being Plato's teacher, the drinker of the hemlock, the author of the theory of Ideas, etc., for such attributes might be used in identifying a given individual as Socrates. The intension of the general term 'red' would be the simple attribute of redness by which we identify objects as being red. The distinction between a term as a linguistic expression and the attributes or relations it expresses for someone who understands it as its intension corresponds to the distinction made in Section 3 between a sentence and a proposition. There can be a variety of terms either of different languages or the same language having the same intension, e.g. 'man', 'homme', and 'human being'.

Singular terms can be distinguished into three basic types. The first of these consists of proper nouns such as 'Socrates', 'Boston', 'Plymouth Rock', 'World War I' naming a particular person, place, thing, or event. The second consists of descriptive phrases containing one or more general terms. 'The most famous teacher of Plato', 'the capitol of Massachussetts', 'the site of the Pilgrim's initial landing', and 'the first international war' would be such phrases in which the object denoted is described in a way that the term can be ascribed to it alone. These descriptive phrases of the form 'the so-and-so' denoting a unique object are called <u>definite descriptions</u>. From the examples just given it should be clear that proper nouns and descriptive phrases can be used interchangeably to denote the same object, that Socrates can be denoted by either 'Socrates' or 'the most famous teacher of Plato'. In inference 2) we encountered an example of how different terms with the same extension can be alternately used while still preserving the connection upon which the validity of an inference depends. It is possible to list as a third type of singular term those consisting of a demonstrative, a pronoun, or a demonstrative qualifying a general term, e.g. 'this', 'he', 'this table', 'that man'. These are typically used when the context together with the constituent indicator word suffices to indicate what the singular term denotes.

General terms can be distinguished into two types, those that denote kinds of objects, e.g. 'man' and 'table', and those that only express attributes of relations that are true of these objects, e.g. 'red', 'wise', and 'hits'. The former are called sortal terms, or simply <u>sortals</u>, the latter <u>attributives</u> The sortal-attributive distinction can be seen to roughly correspond to the grammatical distinction between nouns on the one hand and verbs and adjectives on the other.[3] It would be a mistake, however, to identify general terms as logical elements with words that constitute the vocabulary of a given language. Logical terms are the recurring elements in a certain class of inferences, and there is no necessity that these recurring elements be single words. Thus, 'man' is a sortal general term, but so is the phrase 'very large man with a suitcase'; 'walks' is an attributive, but so also is 'walks quickly while carrying a stick'. The only requirement for a general term is that it be capable of being ascribed to a plurality of objects, and this is satisfied by complexes of words as well as by single words.

When the extension of a term is included in that of another it is said to be <u>subordinate</u> to that other term. This relation of subordination can hold between either sortals or attributives. The substantive 'man' is subordinate, for example, to 'animal', since the class of men is included (is a sub-class of) the class of animals. Similarly, the attributive 'red' is subordinate to 'colored', since red things constitutes a sub-class of things that are colored. Hierarchies of subordination can be found between terms. 'American' is subordinate to 'man', which is subordinate to 'animal'. 'Animal' is in turn subordinate to 'living thing' or 'organism', which is in turn subordinate to 'material thing'. In the same manner 'dark red' is subordinate to 'red', while 'red' is subordinate to 'colored'. 'Colored' is in turn subordinate to the 'quality', since anything that is colored has a quality. Since their extension is only the single individual they denote, singular terms are at the lowest rank in such hierarchies and have no other term subordinate to them. Those terms of highest ranks, perhaps terms like 'material thing', 'number' or 'quality' that are subordinate to no other terms, we shall call **category terms**. They play a very central role in philosophic discussion. The question whether there is some one term such as 'object' or 'entity' to which even these terms are subordinate is a central one for logic and philosophy.

Subjects and Predicates. Our discussion up to now has concentrated on terms apart from the context of sentences in which they may occur. While useful for certain purposes, this is an artificial abstraction. A term is never used by itself for purposes of communication. Instead, the units of communication are sentences, and terms must finally be considered according to their function within a sentence.

The function of the **subject** of a sentence is to denote the object or objects the sentence is about; that of the sentence's **predicate** is to express some attribute or relation that may be true of the denoted objects. If the attribute or relation expressed by the predicate of a sentence is recognized by someone as true of the object or objects denoted by the subject, then he judges the proposition expressed by the sentence as true; if it is recognized as false of the denoted objects, then the proposition is judged false. The quantity of the subject terms of a sentence determines the quantity of the entire sentence. A **singular sentence** is one in which all subjects are singular terms, e.g. 'Socrates is a man', 'The first President of the U. S. was brave', or 'This book is red' whose subjects are respectively 'Socrates', 'the first President of the U. S.', and 'this book'. A **general sentence** is one in which at least one subject is a general term modified by a **quantifier** indicating the quantity of objects denoted by the subject to which the predicate is to be ascribed. Thus, 'All men are mortal', 'Most dogs are tame', and 'Six boys ran swiftly' are general sentences, since their subjects are the general terms 'men', 'dogs', and 'boys' modified by the quantifiers 'all', 'most', and 'six'. Occasionally the quantifier does not explicitly occur in a **general sentence**, as in 'Snakes are dangerous', where it is implicitly understood that the predicate is to be ascribed to all the objects denoted by 'snakes'. Notice that while general terms can function as either subjects or predicates, a singular term can only perform the role of a subject and never that of a predicate. Notice also that while a sortal such as 'man' can function as either a subject as in 'All men are mortal' or a predicate as in 'Some Greeks are men', attributives can function only as predicates. Thus, we say 'Some men are wise', with the sortal 'men' as subject, but not 'Some wise are men'.

There is a more or less exact correspondence between terms considered in isolation and considered as the subjects of sentences. For example, in 'Socrates is wise' and 'Some men are wise' we find the singular term 'Socrates' and the general term 'men' as subjects. This same type of correspondence is not found for predicates of sentences. A sentence's predicate is considered to be the remainder of the sentence after the subjects have been deleted. The predicate of 'Socrates is wise' is '...is wise', the remainder after the deletion of 'Socrates', while that of the general sentence 'Some men were attacked' is '...were attacked'. The independent general term 'wise' becomes in the former an adjective plus the appropriate form of the verb 'to be'; in the latter the term 'to attack' becomes transformed in order to indicate the passive voice and past tense. In the process of being combined with some other term to function as a predicate a given term may be subjected to various grammatical transformations by which the mood, voice, and tense of the sentence is indicated and the verb or adjective made to agree in number with the subject. What these transformations are will be dictated by grammatical rules that will vary from language to language. A predicate is, then, a general term, but a term whose form may be quite different from that which it takes when considered independently of a sentence's context.

In the logical analysis of a sentence into terms there is no requirement that a sentence have but one subject. In 'Cain killed Abel' there are two subjects, since both 'Cain' and 'Abel' denote the individuals the sentence is about. The predicate is '...killed...', which expresses a relation true of the denoted individuals. In 'The largest city in the U. S. lies between Boston and Washington, D. C.' the subjects are the definite description 'the largest city in the U. S.' and 'Boston' and 'Washington, D. C.', and the predicate is '...lies between ...and...' expressing a relation true of the three cities. Indeed, there is no limit to the number of subjects that can at least be theoretically found in a given sentence. Sentences such as these with more than one subject are _relational sentences_. Their constituent predicates are _relational predicates_ in contrast to the _monadic_ or one-place predicates that express attributes that are true of objects denoted by a single subject. Relational sentences can also be general sentences, as for 'Every woman is loved by some man', with the general terms 'woman' and 'man' as its subjects and '...is loved by...' as its dyadic or two-place predicate. The existence of relational sentences makes it obvious that a _logical_ subject of a sentence cannot be necessarily equated with its _grammatical_ subject. In 'Cain killed Abel', for example, the grammatical subject is 'Cain', with 'Abel' the direct object and part of the predicate. A term is assigned as a logical subject by virtue of its denoting role within a sentence. This role can be performed by any noun or noun phrase, whether a grammatical subject or a direct or indirect object. Whatever the rationale of the grammarian's distinction between subject and predicate, it is not that of the logician.

17. Logical Representation of Analyzed Sentences

We shall concentrate our attention in this section and the next on representing the logical form only of sentences in which occur monadic predicates and a single subject. In Section 19 the representation of relational sentences will be discussed.

Atomic Schemata. Arbitrary predicates of the object language are represented in the meta-language of predicate logic by the predicate schemata P,Q,R,...; subjects that are singular terms are represented by the individual schemata a,b,c,... . Singular sentences are represented by juxtaposing the individual schema representing the subject to the right of the predicate schema representing the predicate. Thus, if P is taken as representing the predicate '...is wise' and a the singular term 'Socrates', then the sentence 'Socrates is wise' would be represented by the expression Pa. 'The successor of 4 is a prime number' would be represented by Qb if Q represents the predicate '...is a prime number' and b the definite description 'the successor of 4'. Expressions such as Pa and Qb representing singular sentences analyzed into subject and predicate terms are the atomic sentence schemata of predicate logic. These atomic schemata can be combined or operated on by means of sentence connectives and parentheses as in sentence logic to form molecular schemata representing molecular sentences. For example, Pa ∧ (Qb ⊃ Pb) could represent 'Socrates is wise, and if Plato is a philosopher he is also wise'.

Open Schemata and Variables. Recall that the predicate of a sentence is obtained by deleting the sentence's subjects. Instead now of simply deleting the subjects let us indicate the place they occupy by means of a blank. Thus, instead of the predicates '...is wise' and '...killed...' we write '_____ is wise' and '_____ killed _____', where the blanks indicate the positions to be filled by a subject. These expressions are called open sentences. An open sentence such as '___ is wise' becomes a closed or complete sentence expressing a true or false proposition when a singular term is substituted for the blank. When 'Socrates' is substituted, we have a sentence expressing a true proposition;

when 'Alcibiades' is substituted, an expression of a false proposition. We shall represent open sentences in our meta-language by <u>open sentence schemata</u> in which <u>individual variables</u> x,y,z,... represent the blank positions of subjects. Thus, if P again represents '...is wise', the open sentence '___is wise' is represented by the open schema Px, where x is the individual variable representing the blank. An atomic schema representing a singular sentence is obtained from the open schema by substituting for the individual variable an individual schema. By substituting a for the variable in Px we obtain the schema Pa. If a stands for 'Socrates', we are now representing the sentence 'Socrates is wise'. The individual schema a is called a <u>substitution instance</u> of the variable x.

We may also regard variables as standing for pronouns such as 'he', 'she', or 'it' that occur in sentences. The open schema Px can thus be regarded as representing 'he is wise', where it is left indefinite whom 'he' refers to. Pronouns as they occur in sentences secure their reference in terms of previous information. Thus, if we say 'Socrates is a philosopher and he is wise', we understand the pronoun 'he' to refer to Socrates, since the first sentence of the conjunction informs us that it stands in place of the name 'Socrates'. But if we abstract from such previous information and leave reference indefinite, we have pronouns that can be represented by the variables of open schemata.

<u>Quantificational Schemata</u>. We represent general sentences by means of open schemata prefixed by logical constants called <u>quantifiers</u>. There are a variety of words that serve as quantifiers in natural languages, words such as 'all', 'some', 'a few', 'many', 'most' and the numerical quantifiers 'one', 'two', etc. Only two of these quantifiers are represented by special symbols in predicate logic. The quantifier 'all' (or 'every') is represented by the <u>universal quantifier</u> $\forall x$; the quantifier 'some' ('there is at least one' or 'there exists') is represented by the <u>existential quantifier</u> $\exists x$. Other quantifiers are either ignored or defined in terms of these two.

The representation of general sentences involves a series of paraphrases in which pronouns are introduced and replaced by variables. Consider the sentence,

    1) Everything is extended.

This can be paraphrased as

    2) For every thing it is extended
or 3) Everything is such that it is extended.

Let us now replace the subject 'thing' of 3) by the variable x and indicate that the pronoun 'it' refers back to this subject by using the same variable to represent it also. Then we have the sentence,

    4) Every x is such that x is extended.

By introducing the universal quantifier $\forall x$ to represent 'every x' and the predicate schema P to represent '...is extended' 4) becomes represented by

    5) $\forall x P x$

Sentences of universal quantity such as 1) and its paraphrases 2)-4) are called <u>universal sentences</u> and are represented by <u>universal quantificational schemata</u> such as 5).

Note that the role of the variables in 4) and 5) is that of allowing the cross reference performed in 2) and 3) by the pronoun 'it'. The sentence is about all things; the variables serve to indicate that it is these things that are what is extended. Note also that the subject of sentence 1), the word 'thing', becomes represented by the variable within the quantifier $\forall x$. In general, variables within quantifiers will represent subjects of general sentences, with later occurrences of the variable referring back to this subject. The sentence,

    6) All men are mortal

can in the same way be paraphrased as

    7) All men are such that they are mortal.

The variable x can again be introduced to represent the sentence's subject 'men'. This produces

    8) All xs are such that xs are mortal.

This expression would now be represented by

    9) $\forall x Q x$

with Q representing the predicate '...are mortal'. As we shall see in the next section, however, the more usual paraphrase of 6) in predicate logic would involve introducing a more general subject term.

The class of objects denoted by the variable within the quantifier representing a sentence's subject is said to constitute the <u>domain</u> (or universe of discourse) of the variable. The domain for <u>5) is</u> thus the class of (material) things, since the variable is understood as representing the subject 'thing'. The domain for 9) is the class of men, since this class is the extension of 'men' which the variable in the quantifier is taken to represent. The function of the universal quantifier ∀x in 9) is to indicate it is all of the members of the domain constituted by the class of men that the predicate represented by Q is to be ascribed to.

A representation similar to that just outlined can be given for existential sentences, sentences with an existential quantifier. For example, the sentence,

    10) Something is heavy

can be paraphrased as

    11) There is some thing such that it is heavy.
or 12) There is (exists) at least one thing such that it is heavy.

With variables representing the subject 'thing' and the pronoun 'it' 12) becomes

    13) There is at least one x such that x is heavy.

With ∃x for 'at least one x' and Px for 'x is heavy', 12) is finally represented by

    14) ∃xPx

The domain denoted by the variable would be the class of things, since 'thing' is the subject of 10). The existential quantifier indicates that it is at least one member of this domain to which the predicate represented by P is to be ascribed.

By means of the negation sign various negative forms of universal and existential sentences can be represented. A sentence of the form 'Nothing is P' (or 'No thing is such that it is P') is represented by ∀x-Px, while 'Something is not P' becomes represented by ∃x-Px. This latter is the contradictory of ∀xPx, since to deny that everything is P is to affirm that something is not P. ∀x-Px is in the same way the contradictory of ∃xPx, since nothing is P if and only if it is not the case that something is P. The basic logical relations between these four forms of sentences can be summarized in the following so-called "Square of Opposition":

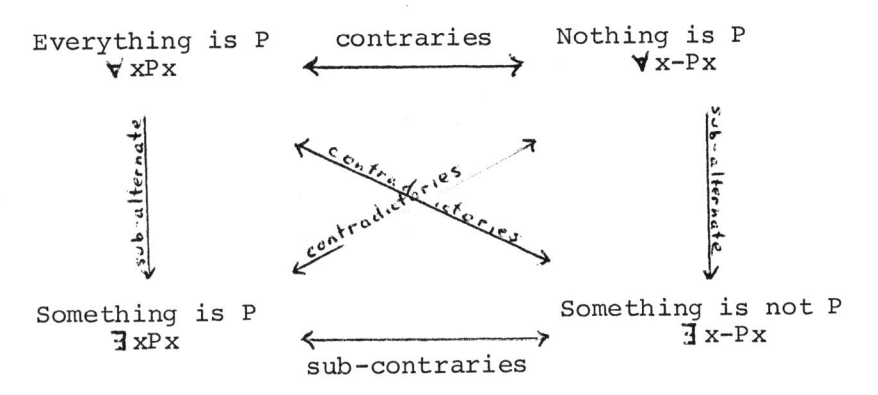

```
Everything is P      contraries        Nothing is P
   ∀xPx          <─────────────>         ∀x-Px
       │   ↖            ↗   │
       │     ↖ contradictories ↗   │
sub-   │       ↘        ↗       │  sub-
alternate│         ╳            │ alternate
       │       ↗        ↘       │
       │     ↗            ↘     │
       ↓   ↙                ↘   ↓
 Something is P                    Something is not P
   ∃xPx          <─────────────>         ∃x-Px
                 sub-contraries
```

Two propositions are said to be <u>contraries</u> of one another when it is impossible for both to be true at the same time, though both may be false. This relation holds between the propositions expressed by ∀xPx and ∀x-Px. It is false that all things are red and false that all things are not red, since things are of different colors. But it is clearly impossible for it to be true that everything is red and at the same time it be true that everything is not red. Two propositions are related as <u>sub-contraries</u> when both cannot be false. This relation holds for ∃xPx and ∃x-Px as indicated in the diagram, if we assume the domain denoted by the variable to be non-empty. Some things are red and some things are not red, but if there is at least one thing to which a color word like 'red' is applicable in our domain, both propositions could not be false. Two propositions are contradictory if they must have opposite truth values. The relations of contradiction can be summarized by two important logical equivalences:

$$\forall xPx \Leftrightarrow -\exists x-Px$$
$$\exists xPx \Leftrightarrow -\forall x-Px$$

Existential propositions expressed by sentences of the form ∃xPx and ∃x-Px are said to be <u>sub-alternates</u> of the universal propositions ∀xPx and ∀x-Px that entail them. That these universal propositions do entail the existential seems obvious enough. For it to be true that everything is extended, there must be things to which we ascribe the attribute of being extended. Hence, that at least one thing is extended must also be true. Similar considerations hold for the negative forms.

   The assumption that there exists at least one member of the domain denoted by the variable is usually considered essential to predicate logic. If the domain were empty and contained no members, then it seems it would be impossible to say anything with reference to it. Without an object to be denoted, no proposition could be expressed that is either true or false. The assumption of a non-empty domain should not be understood as committing us to domains consisting of what can be empirically identifiable, of what exists in the sense that tables and chairs do. Our domain can be chosen to be the rational numbers, points in geometric space, or fictitious entities such as the class of dragons. By assuming that the chosen domain has at least one member we seem to be assuming only that what we say about it is meaningful and either true or false.[1]

Exercises. Represent the logical form of the following sentences, using the suggested symbolism. Also specify the domain denoted by the variable where the represented sentence is general.
1. Nothing is eternal. (E: '...is eternal')
2. Not everything is eternal.
3. Something is valuable. (V: '...is valuable')
4. There is nothing that is valuable.
5. Some numbers are not odd. (O: '...is odd')
6. Not every event is caused. (C: '...is caused')
7. Lions are dangerous. (D: '...is dangerous')
8. No lions are dangerous.
9. Lions exist. (L: '...is a lion')
10. Centaurs do not exist. (C: '...is a centaur')
11. Lions are in the cage. (C: '...is in the cage')
12. No one likes to lose. (L: '...likes to lose')
13. Tom is both hungry and tired. (t: 'Tom'; H: '...is hungry'; T: '...is tired')
14. Neither Tom nor Harry is hungry. (h: 'Harry')
15. Tom is hungry, but someone is not tired.

18. Complex Quantificational Schemata

So far we have restricted ourselves to representing general sentences in which occur a single predicate. With the help of the connectives of sentence logic it is possible to represent increasingly more complex sentences.

Introduction of General Subjects. We have seen how

1) All men are mortal

can be represented by $\forall x Q x$ if we understand the variable in the quantifier to represent the subject 'men' and its domain to be the class of men. But by introducing a more general subject, a subject with a wider extension, it is possible to represent 1) in a different way. The following paraphrase in which we introduce as subject the word 'thing' seems to express the same proposition as 1):

> 2) Everything is such that if it is a man then it is mortal.

Using occurrences of the variable x to represent the subject 'thing' and the occurrences of the pronoun 'it' that refer back to this subject, 2) becomes

> 3) Every x is such that if x is a man, then x is mortal.

Since 'if...then' is represented by the material implication sign and 'every x' by the universal quantifier, 3) can be represented by

> 4) $\forall x(x \text{ is a man} \supset x \text{ is mortal})$.

Introducing the predicate schema P for '...is a man' and Q for '...is mortal', our sentence can now be represented by

> 5) $\forall x(Px \supset Qx)$

It is important that the material implication sign be used in representing 1). It would not be correct, for example, to represent this sentence by $\forall x(Px \wedge Qx)$. In saying that all men are mortal we are clearly not saying that everything is both a man and mortal. We are instead saying that for everything in our domain of things if it is a man then it is mortal, or alternatively, for anything we choose to select from this domain it will be mortal if it turns out to be a man.

By paraphrasing 1) by 2) we have introduced as a more general subject the term 'thing' and converted the former subject 'men' into the predicate '...is a man'. As a result the domain changes from the class of men to the class of things. As the antecedent of the conditional 'if it is a man then it is mortal' the predicate '...is a man' serves to specify which of the members of this enlarged domain the predicate '...is mortal' is to be ascribed. It is usually the practice to introduce as subject terms in such paraphrases as 2) substantive category terms such as 'thing', 'event' or 'number'. By doing this we insure a common subject for all the sentences of the inference we are evaluating. The sentence 'No odd number is divisible by 2' would thus be typically paraphrased as 'Every number is such that if it is odd then it is not divisible by 2' in which 'number' replaces 'odd number' as the subject. Its representation becomes $\forall x(Px \supset -Qx)$. Whether there is a single category term such as 'object' or 'entity' that can be introduced for every sentence is a subject of controversy. If we were to permit such an introduction, then 'Every number is either odd or even' could be paraphrased by 'Every object is such that if it is a number it is either odd or even'. In the paraphrase 'number' becomes converted from a subject to the predicate '...is a number'. The domain denoted by the variables used to represent the paraphrase would be class of objects in general.

Similar considerations hold for existential sentences. We previously would have represented

   6) Some men are wise

by $\exists x Qx$ by restricting our domain to the class of men. But we can also introduce again the category term 'thing' as subject and paraphrase 6) by

   7) There is at least one thing such that it is a man and it is wise.

Introducing the existential quantifier, variables, and conjunction sign, 7) becomes

   8) $\exists x(x \text{ is a man} \wedge x \text{ is wise})$.

With P for '...is a man' and Q for '...is wise', the logical representation of 7) and hence 6) is

   9) $\exists x(Px \wedge Qx)$

It is important that the conjunction sign be used in the representation of 6). We could not represent it by $\exists x(Px \supset Qx)$, for example, since 'There is at least one thing such that it is wise if it is a man' would not be a correct paraphrase. This latter sentence does not commit us to saying that there is a man who is also wise as does our original. As did the antecedent of the open schema within 5), the first of the conjuncts of $Px \wedge Qx$ in 9) represents the open sentence containing the predicate specifying which of the class of things that constitutes the enlarged domain the predicate represented by Q is being ascribed to. It is things that are men which our sentence claims are wise.

  <u>Bound and Free Variables</u>. Notice in 5) and 9) above the use of parentheses. They serve to indicate that occurrences of the variable x refer back to prefixed quantifiers. In $\forall x(Px \supset Qx)$, for example, the xs such that if they are P they are Q are the xs quantified by $\forall x$, and this is shown by enclosing the open schema $Px \supset Qx$ within parentheses. Parentheses in this manner indicate what is called the <u>scope</u> of the quantifier, the open schema to which a prefixed quantifier applies and whose variables refer back to the variable in the quantifier. The parentheses of $\forall x(Px \supset Qx)$ thus serve to indicate that the schema $Px \supset Qx$ lies within the scope of the universal quantifier.

Those occurrences of a variable within the scope of a quantifier which contains that variable are said to be <u>bound variables</u>, variables bound by the prefixed quantifier. Any occurrence of a variable not bound by a quantifier is said to be a <u>free variable</u>. By indicating the scope of a quantifier parentheses thus indicate which variables are bound and which free. In $\forall x(Px \supset Qx)$, for example, both occurrences of x in $Px \supset Qx$ are bound variables. In $\forall xPx \supset Qx$ where the scope of the quantifier extends only to Px (we omit parentheses around Px for the sake of economy), the x in Qx is free. Similarly, in $\exists x(Px \wedge Qx)$ both occurrences of x in $Px \wedge Qx$ are bound, while in $\exists xPx \wedge Qx$ the x in Qx is a free variable outside the scope of the existential quantifier.

Expressions containing free variables are open schemata representing open sentences that fail to express a proposition. We used such expressions in the previous section when we took Px to represent '___ is mortal' or 'he is mortal' with the reference of the pronoun left indefinite. In similar fashion $Px \supset Qx$ might represent the open sentence 'if ___ is a man then ___ is mortal' or 'if it is a man then it is mortal'. Schemata containing no free variables and representing complete sentences are <u>closed schemata</u>. An open schema can be converted to a closed schema, as we have seen, by either substituting an individual schema for the free variables or binding the variables by a prefixed quantifier. Thus, the open schema $Px \wedge Qx$ could become $Pa \wedge Qa$ if we substituted for both occurrences of the variable; or it could become $Pa \wedge \forall xQx$ if we substituted for the first and bound the second.

By extending the scope of quantifiers quantificational schemata can be formed that represent increasingly complex general sentences. A sentence such as 'All happy men are both wise and fortunate', for example, may be paraphrased as 'Everything is such that if it is happy and it is a man then it is wise and it is fortunate'. It would then be represented by the schema $\forall x(Px \wedge Qx \supset Rx \wedge Sx)$, with P for '...is happy', Q for '...is a man', R for '...is wise' and S for '...is fortunate'. Notice that we continue to use the binding conventions of sentence logic for the open schema $Px \wedge Qx \supset Rx \wedge Sx$ within the scope of the quantifier. Quantificational schemata can in turn be combined or operated on by the sentence connectives in order to represent still more complex sentences. Thus, $\forall x(Px \supset Qx) \wedge \exists x(Px \wedge -Rx)$ might be used to represent 'All men are mortal and some men are not wise'.

The underlying logical structure of a sentence can often be disguised by the occurrence of logical words in its surface structure. The word 'any', for example, is usually synonymous with 'every' and represented by the universal quantifier. The sentence 'Any person who wins the lottery will be rich' would clearly be represented by $\forall x(Px \land Qx \supset Rx)$, with P for '...is a person', Q for '...wins the lottery', and R for '...will be rich'. But in certain sentence contexts 'any' can be represented by the existential quantifier. Thus the sentence 'If any person wins the lottery someone will be rich' would be represented by the schema $\exists x(Px \land Qx) \supset \exists x(Px \land Rx)$, since the antecedent of the conditional only assumes that at least one person wins the lottery. Even the logical meaning of 'some' can vary with context. Almost invariably it is represented by the existential quantifier. But as occurring in the sentence 'If some man wins the lottery he will be rich' it would be represented by the universal quantifier. This sentence is synonymous with 'Any man who wins the lottery will be rich' and is thus to be represented, as we have seen, by $\forall x(Px \land Qx \supset Rx)$. The logical word 'and' can also cause difficulties. It would not be correct to represent 'All animals and plants live either on the land or in water' by $\forall x(Px \land Qx \supset Rx \lor Sx)$, since the sentence does not state that everything that is both an animal and a plant lives on land or in water. The correct representation is rather $\forall x(Px \lor Qx \supset Rx \lor Sx)$ or the logically equivalent $\forall x(Px \supset Rx \lor Sx) \land \forall x(Qx \supset Rx \lor Sx)$.

It must be pointed out that there are usually many alternative ways of representing the form of a given sentence, and which we choose will often depend on the inference context in which it occurs. For example, the sentence 'Everyone is happy or tired' would be represented in different ways as occurring in the following inferences:

10)   Everyone is happy or tired.
      John is a person.
      ∴ John is happy or tired.

11)   Everyone is happy or tired.
      No one is happy.
      ∴ Everyone is tired.

The first premiss of 10) could most simply be represented as $\forall x(Px \supset Qx)$, with P for '...is a person', Q for '...is happy or tired', and the variable x for the category term 'thing'. As a premiss of 11), however, we would represent it by $\forall x(Qx \lor Rx)$, with Q now for '...is happy', R for '...is tired', and x for the general term 'person'. The selection of different representations is due to a difference in the common subject of the inferences and the recurring predicate terms. In 10) 'person' occurs within a predicate in the second premiss and we must therefore introduce the more general term 'thing' as the common subject of the first premiss and conclusion. The predicate '...is happy or tired' recurs in the first premiss and conclusion. In 11), on the other hand, the common subject is 'person', and the recurring predicates are '...is happy' and '...is tired'. We usually introduce category terms as subjects of the paraphrases of general sentences because this insures a common subject for the inferences in which they occur. But in certain inferences such as 11) a simpler representation is available by retaining the subject of the original sentence.

Exercises. Represent the logical form of the following sentences by introducing category terms as subjects.
1. Some lions are not dangerous. (L: '...is a lion'; D: '...is dangerous')
2. All lions are either tame or dangerous. (T: '...is tame')
3. No lion which is tame is dangerous.
4. A lion is dangerous if it is not tame.
5. Only lions are dangerous.
6. Only tame lions are not dangerous.
7. No lion is dangerous unless it is untame.
8. Some lions are dangerous unless they are tame.
9. No prime number is odd. (P: '...is prime'; O: '...is odd')
10. Not every unusual event is a miracle. (U: '...is unusual'; M: '...is a miracle')
11. Someone failed the test although he studied. (P: '...is a person'; F: '...failed the test'; S: '...studied')
12. Someone studied, and not everyone failed the test.
13. All plants and animals require oxygen and light. (P: '...is a plant'; A: '...is an animal'; O: '...requires oxygen'; L: '...requires light')
14. Only plants and animals require oxygen and light.
15. Anyone will succeed if he is tenacious. (P: '...is a person'; S: '...will succeed'; T: '...is tenacious')
16. Someone will succeed only if he is tenacious.
17. If anyone succeeds, then someone is tenacious.
18. One who does not succeed is one who is not tenacious.
19. If a person does not succeed then not everyone succeeds who is tenacious.
20. If everyone succeeds who is tenacious, then Jones will succeed. (j: 'Jones')

19. Representation of Relational Sentences

We now extend the symbolism just introduced to the representation of sentences in which occur relational predicates expressing relations holding between objects denoted by two more subjects.

For singular sentences this representation is relatively straight-forward. We simply juxtapose to the right of the predicate schema representing the relational predicate the individual schemata representing the subjects. Thus, if R represents the dyadic predicate '...killed...', a the singular term 'Cain', and b 'Abel', then 'Cain killed Abel' becomes represented by Rab. The order of the individual schemata reflects that of the singular terms that are represented. A different order to produce Rba would represent the entirely different sentence 'Abel killed Cain'. For a relational sentence with more than two subjects the representation is similar. The representation of 'The large red brick building was constructed between Morris Library and Lawson Hall' would be Sabc, with S representing the three-place or triadic predicate '...was constructed between...and...'. The sentence 'John gave five dollars to Peter for the knife' would be represented by Rabcd, with R representing now the four-place or tetradic predicate '...gave...to...for...'. For a sentence containing any number n of subjects and an n-place or n-adic predicate a similar representation is possible.

For general sentences in which one or more of the subjects are general terms the representation can be more difficult. For this representation we use variables denoting a common domain together with the universal and existential quantifiers. With one of our subjects a general term and the others singular terms the variable occupies a place beside the predicate schema and is bound by a prefixed quantifier. Thus, 'Mary likes everything' would be paraphrased as 'Everything is such that Mary likes it', and with R for '...likes...' would be represented by the quantificational schema $\forall x Rax$. Here 'thing' is the general subject and 'Mary' the singular. 'Something likes Mary' would be represented by $\exists x Rxa$. As is apparent, the position of x and a in the open schema indicates the direction of the relation, whether directed towards all the things or towards Mary.

With more than one general subject we must introduce more than one kind of variable bound by prefixed quantifiers. For example, the representation of 'Everything attracts everything' would be $\forall x \forall y Rxy$ and of 'Something attracts something' $\exists x \exists y Rxy$, with R representing the dyadic predicate '...attracts...'. When both quantifiers are of the same type, it is evident that the order of the quantifiers is indifferent. We have, then, the following equivalences:

$$\forall x \forall y Rxy \Leftrightarrow \forall y \forall x Rxy$$
$$\exists x \exists y Rxy \Leftrightarrow \exists y \exists x Rxy$$

But when we combine universal and existential quantifiers to bind the variables their order is important. To illustrate this, let the open schema Rxy represent the open sentence '___ is less than ___' and let the variables x and y denote the positive integers. For every integer there is some integer that it is less than (its successor, for example), and thus $\forall x \exists y Rxy$ is true. But there is no integer such that every integer is less than it, and $\exists y \forall x Rxy$ is thus false. Hence the two expressions are not equivalent.

In expressions with two or more prefixed quantifiers we are assuming that quantifiers to the left have increasingly wider scope. This assumption allows us to delete parentheses that would otherwise be necessary. For example, $\forall x (\exists y Rxy)$ can be written in the manner of the preceding paragraph as $\forall x \exists y Rxy$, it being understood that the scope of the existential quantifier is Rxy, while that of the universal quantifier $\forall x$ is $\exists y Rxy$. Similarly, instead of $\forall x \{\exists y [\exists z (Px \land Rxyz)]\}$ we can write simply $\forall x \exists y \exists z (Px \land Rxyz)$.

In the examples considered so far the variables have been taken as representing the general subject term of the object language sentence. As before for monadic predicates, it is often necessary to introduce by paraphrase a more general subject and convert the original subjects into predicates. Consider, for example, the sentence,

    1) Every man owns something

whose subjects are the general terms 'man' and 'thing'. This can be paraphrased as

    2) Everything is such that if it is a man then there is something it owns.

If the variable x is taken to represent the category term 'thing' in 'everything' and y the term 'thing' in 'something' and the universal and existential quantifiers introduced, 2) becomes

    3) $\forall x [x \text{ is a man} \supset \exists y (x \text{ owns } y)]$

With P for '...is a man' and R for '...owns...', 3) is finally

    4) $\forall x (Px \supset \exists y Rxy)$

Similarly, the sentence,

    5) Every man loves some woman

with subjects 'man' and 'woman' becomes

> 6) Everthing is such that if it is a man, then there is something which is a woman and loved by it (the man).

It can now be represented by

> 7) $\forall x[Px \supset \exists y(Qy \wedge Rxy)]$

with P for '...is a man', Q for '...is a woman', and R for '...loves...'.

Note that in the paraphrase of 1) by 2) and of 5) by 6) we introduce a single subject 'thing' in place of the original two subjects. In the paraphrase 6) 'man' and 'woman' become converted from subjects to the predicates '...is a man' and '...is a woman' and now serve to specify what kinds of things the relation of loving holds between. When the quantifier is universal the specification is by means of an antecedent of an implication; when the quantifier is existential it is by a conjunct. This type of paraphrase is a required preliminary for the representation of any sentence with two or more different subjects. First we must substitute for the different subjects a single one, usually by introducing a category term such as 'thing' or 'number'. The subject of the paraphrase will now denote a domain of objects between which the relations expressed by the sentence hold. We then must specify the sub-classes within this domain for which the relations hold. In 5) we saw that we specified that the relation of loving holds between those things that are men and those that are women. The quantifiers serve to indicate the quantity of objects in these sub-classes that are related, e.g. all men to at least one woman, as in 7).

It is often convenient to represent relational sentences by a step-wise procedure in which the major parts are first paraphrased. We might, for example, introduce variables x and y to stand for 'thing', the universal quantifier, and connectives, and paraphrase 5) first by

> 8) $\forall x(x \text{ is a man} \supset x \text{ loves some woman})$

The consequent, 'x loves some woman', can now be paraphrased as

> 9) $\exists y(y \text{ is a woman} \wedge x \text{ loves } y)$

Introducing the predicate schemata P, Q, and R as before we represent the result of replacing the consequent of the implication in 8) by 9). This now produces the schema 7).

For more complicated relational sentences this procedure is especially convenient. Consider, for example, the sentence,

10) No man gives everything he owns to some woman he loves.

This is of the general form of the universal sentence,

11) ∀x[x is a man ⊃ -(x gives everything he owns to some woman he loves)

The consequent within 11) becomes

12) -∀y(x owns y ⊃ (x gives y to some woman he loves)

The consequent within 12) becomes in turn

13) ∃z(z is a woman ∧ x loves z ∧ x gives y to z)

Representing the predicate '...is a man' by P, '...is a woman' by Q, the dyadic predicates '...owns...' by R and '...loves...' by S, and the triadic predicate '...gives...to..' by T, 11), 12), and 13) can be combined and represented by

14) ∀x{Px ⊃ -∀y[Rxy ⊃ ∃z(Qz ∧ Sxz ∧ Txyz)]}

14) is now the correct representation of 10). Without this step-wise procedure the representation would be extremely difficult.[1]

The use of different variables and parentheses in schemata such as 7) and 14) enables us to indicate which variables are bound by the prefixed quantifiers. A variable is bound by a quantifier containing that variable provided it lies within its scope. In ∀x[Px ⊃ ∃y(Qy ∧ Rxy)] the outer brackets indicate that the scope of the universal quantifier includes the open schema Px ⊃ ∃y(Qy ∧ Rxy). The occurrences of x in Px and Rxy are thus shown to be bound by the universal quantifier, since it contains that variable. Similarly, the inner parentheses show that the scope of the existential quantifier includes the open schema Qy ∧ Rxy. Since the existential quantifier includes the variable y, the occurrences of y in Qy and Rxy are bound by it. A useful way of indicating which quantifiers bind which variables is to connect by means of lines the quantifiers to the variables they bind. Thus, 7) can be diagrammed as

∀x[Px ⊃ ∃y(Qy ∧ Rxy)]

14) can likewise be diagrammed as

∀x{Px ⊃ -∀y[Rxy ⊃ ∃z(Qz ∧ Sxz ∧ Txyz)]}

Here the braces indicate the scope of the universal quantifier $\forall x$. The brackets indicate the scope of $\forall y$, while the parentheses indicate the scope of $\exists z$.

Again, it must be emphasized that the logical representation of a sentence will depend on the inference context in which it occurs. In general, a relational predicate must be represented as such only if recurrences of it are essential for the validity of the inference in which it occurs. For example, as the premiss of the inference,

> Bill is older than Tom.
> ∴ Tom is not older than Bill.

'Bill is older than Tom' must be represented by Rab, with R for '...is older than...', since recurrences of this relational predicate is essential for evaluating the inference valid. In contrast, within the inference,

> Bill is older than Tom.
> ∴ Someone is older than Tom.

the sentence could be represented by Pa, with P for the monadic predicate '...is older than Tom'. Here the validity of the inference depends only on recurrences of the expression '...is older than Tom' taken as a whole.

Exercises. Represent the logical form of the following relational sentences.
1. There is something that attracts everything. (A: '...attracts...')
2. Everything attracts something.
3. John owns something. (j: 'John'; O: '...owns...')
4. Every number has a successor. (S: '...is the successor of...')
5. Not every man loves a woman. (M: '...is a man'; W: '...is a woman'; L: '...loves...')
6. No man loves every woman.
7. Some woman loves some man.
8. There are some odd numbers that are less than some even numbers. (O: '...is odd'; E: '...is even'; L: '...is less than...')
9. No even number is less than every odd number.
10. No odd number is divisible by 2. (a: '2'; D: '...is divisible by...')
11. Anyone who is rich owns a yacht. (P: '...is a person'; R: '...is rich'; Y: '...is a yacht'; O: '...owns...')
12. Anyone who owns a yacht is rich.
13. Only those who own yachts are rich.
14. No one is rich unless he owns a yacht.

15. Everyone is rich unless he owns a yacht.
16. If someone owns a yacht then he is rich.
17. If someone owns a yacht then someone is rich.
18. If any man buys a tie at a store, then there is someone who earns some cash. (M: '...is a man'; T: '...is a tie'; S: '...is a store'; P: '...is a person'; C: '...is cash'; B: '...buys...at...'; E: '...earns...')
19. If any man buys a tie at a store then he earns some cash.
20. No man buys a tie at a store unless he earns some cash.

NOTES FOR CHAPTER IV

Section 16.

1. For the historical development of this branch of logic from the syllogistic logic of Aristotle to the quantificational logic developed principally by Peirce and Frege see again the Kneales [Development] and Bochenski [History].

2. We use Mill's term "denotation" for the relation between the subject of a sentence and the objects it stands for when the sentence is being considered as a constituent of an inference. This is a dyadic relation between term and objects abstracted from any user of the sentence. Outside such logical contexts where the sentence is used by speakers to make statements this relation is commonly referred to the subject's "reference." Here it is considered a triadic relation between the subject term and objects relative to a speaker. For a discussion of the subject-predicate distinction in terms of the referring-ascribing functions see Strawson [Individuals], Chs. V and VI.

3. For the basis of the sortal-attributive distinction (or between what he terms "sortal universals" and "characterizing universals") see Strawson [Individuals], pp. 168-175.

4. Category terms are called "universal words" by Carnap in [Logical Syntax], Sec. 16. In [Empiricism], he maintains that there are separate language frameworks for material things, numbers, etc., and that there is no one framework inclusive of the rest. Since each language framework requires a different domain for the variables used to represent sentences within it, there is in Carnap's view no single category term to which all others are subordinate. For the opposing view see Quine [Paradox], Essay 11.

Section 17.

1. For a discussion of the necessity of a non-empty domain see Cohen [Diversity], pp. 255-262 (reprinted as Selection 15 in Iseminger [Logic]). See also the Kneales [Development], pp. 706, 707 for an argument for its necessity.

Section 19.

1. This procedure is due to Copi [Logic], Ch. V.

V. DECISION PROCEDURES FOR PREDICATE LOGIC

20. Decision Procedure for Finite Domains

We turn now to the problem of deciding for any given inference with recurring terms with the same content whether it is valid or invalid and to the more general problem of deciding whether sentences of this form express tautologies or not. We shall concentrate in the first four sections on inferences in which occur only monadic predicates. In Section 24 the decision problem for those in which relational predicates occur will be discussed.

Evaluation of Inferences. When the domain denoted by the common subject of the sentences within an inference consists of a finite number of individuals, we can construct a decision procedure using the methods of sentence logic. Suppose we take as our domain two colored objects labelled by the terms 'No. 1' and 'No. 2'. It is evident that to say that all of these objects are red is to say that No. 1 is red and that No. 2 is red. The universal sentence 'Everything is red' is thus logically equivalent to the conjunction of singular sentences 'No. 1 is red and No. 2 is red'. Representing '... is red' by the predicate schema P, 'No. 1' by the individual schema a, and 'No. 2' by b, we have then the equivalence

$$\forall x Px \Leftrightarrow Pa \wedge Pb$$

It is evident also that the equivalence between the universal quantificational schema and a conjunction of atomic schemata is not changed if we add to our domain additional objects. If we have n number of objects in our domain names of which are represented by the individual schemata $a_1, a_2, \ldots, a_n$, then we have

1) $\quad \forall x Px \Leftrightarrow Pa_1 \wedge Pa_2 \wedge \ldots \wedge Pa_n$

For the existential sentence 'Something is red' we would have as equivalent the disjunction 'No. 1 is red or No. 2 is red', since the existential sentence will be true if and only if at least one of these disjuncts is true. And in general, for any domain with n objects each denoted by a singular term we have the equivalence

2) $\quad \exists x Px \Leftrightarrow Pa_1 \vee Pa_2 \vee \ldots \vee Pa_n$

For any universal and existential sentence similar equivalences can be stated. We would have as equivalent to the universal sentence 'All the balls are red' relative to our domain of two objects the conjunction 'If No. 1 is a ball then it is red and if No. 2 is a ball then it is red also'. And generally for any sentence of the form 'All P is Q' relative to a domain of n objects we have

3) $\quad \forall x(Px \supset Qx) \Leftrightarrow (Pa_1 \supset Qa_1) \wedge (Pa_2 \supset Qa_2) \wedge \ldots \wedge (Pa_n \supset Qa_n)$

Similarly, for the existential form 'Some P is Q' we have

4) $\quad \exists x(Px \wedge Qx) \Leftrightarrow (Pa_1 \wedge Qa_1) \vee (Pa_2 \wedge Qa_2) \vee \ldots \vee (Pa_n \wedge Qa_n)$

Equivalences 1) and 2) can be seen to allow us to state any general sentence in this manner as a truth function of atomic sentences.

To see how these equivalences can be used in the evaluation of inferences, consider the inference,

<u>Everything is red.</u>
∴ All balls are red.

where the subject 'thing' in the premiss is understood to denote our domain of two colored objects. Using the equivalences 1) and 3) this inference can be represented by

$$\frac{\forall x Qx}{\forall x(Px \supset Qx)} \Leftrightarrow \frac{Qa \wedge Qb}{(Pa \supset Qa) \wedge (Pb \supset Qb)}$$

The inference represented by the right side of this equivalence can now be evaluated by the procedures of sentence logic. The inference would be valid if it is impossible that the premiss of the form $Qa \wedge Qb$ is true while the conclusion $(Pa \supset Qa) \wedge (Pb \supset Qb)$ is false. We can construct a truth table listing the possible interpretations of the atomic constituents of the inference that allows us to determine this.

Qa	Qb	Pa	Pb	Qa∧Qb	(Pa⊃Qa)	∧	(Pb⊃Qb)
T	T	T	T	T	T	T	T
T	T	T	F	T	T	T	T
T	T	F	T	T	T	T	T
T	T	F	F	T	T	T	T
T	F	T	T	F	T	F	F
T	F	T	F	F	T	T	T
T	F	F	T	F	T	F	F
T	F	F	F	F	T	T	T
F	T	T	T	F	F	F	T
F	T	T	F	F	F	F	T
F	T	F	T	F	T	T	T
F	T	F	F	F	T	T	T
F	F	T	T	F	F	F	F
F	F	T	F	F	F	F	T
F	F	F	T	F	T	F	F
F	F	F	F	F	T	T	T

Since for no one of the possible interpretations of the four constituent atomic schamta Pa, Pb, Qa, and Qb is there a true premiss and a false conclusion, the inference is shown to be valid.

This evaluation procedure can be applied to any inference where the domain is understood to be finite. We first represent the inference by quantificational schemata and then replace these schemata by their logically equivalent truth functions of atomic schemata. For an inference with m number of constituent predicates restricted to a domain with n individuals, there will be m x n resulting atomic constituents. We then construct a truth table listing the $2^{m \times n}$ possible interpretations of these constituents. If for no one of these interpretations are the premisses true and the conclusion false the inference is valid for the number of individuals considered; otherwise the inference is invalid. For the inference just evaluated there were 2 predicates restricted to a domain of 2 individuals. Hence the 2 x 2 = 4 atomic constituents and 2^4 = 16 possible interpretations. For five predicates and four individuals there would be 5 x 4 = 20 constituents and 2^{20} possible interpretations. For large numbers of predicates and large domains this method thus results in an extremely lengthy and involved evaluation. Nevertheless, it is in principle possible, and yields a definite answer in every instance.[1]

General Decision Procedure. A general decision procedure applicable to any general sentence can be constructed as an extension of this evaluation procedure for inferences. Since every general sentence whose subject denotes a finite domain is, as we have seen, logically equivalent to some truth function of atomic sentences, its logical status can be assessed by a truth table listing the possible interpretations of these atomic sentences. For example, the sentence 'Everything is red or not red' of the form

5)   ∀x(Px ∨ -Px)

where the domain is restricted to three individuals, is logically equivalent to a sentence of the form

6)   (Pa ∨ -Pa) ∧ (Pb ∨ -Pb) ∧ (Pc ∨ -Pc)

Since there is one predicate and a domain of three individuals, there are 3 x 1 = 3 atomic constituents of which there are altogether 2^3 = 8 possible interpretations. A truth table listing these interpretations would show a sentence of the form of 6) to be true for all of them. Hence the sentence would be shown to be a tautology relative to a domain restricted to three individuals. If a sentence is true for some interpretations but false for others relative to a domain of a certain size, it expresses a contingent proposition relative to that domain; if it is false for all interpretations, it expresses a contradiction.

Notice that this procedure must assume that the domain in question is non-empty. If we take as our domain the chairs in the next room and there are, in fact, no chairs there, then no matter what general sentences we formulate about this domain they will be incapable of being replaced by truth functions of atomic sentences. The procedure also assumes that to every individual in the finite domain we can assign some singular term that denotes it. If such terms are not available for some reason in our object language, then there is again no logically equivalent molecular sentence whose constituents are all atomic sentences.

Infinite Domains. For a domain of infinite size, however, this procedure cannot be directly employed. For then the number of atomic constituents would be infinite: every universal or existential sentence would be equivalent to an infinitely long conjunction or disjunction. It would be therefore impossible to exhaustively enumerate all possible interpretations, since we cannot construct a truth table with infinitely many columns. This is enough to disqualify the procedure as a general method of determining the logical status of a general sentence. And this is more than a theoretical short-coming. General sentences, certainly those typical in science and mathematics, are not about a domain of a certain definite size where we can enumerate all the members. When we say, for example, 'All mammals have hearts' or 'All numbers are odd or even' we are not speaking of a definite and enumerable number of mammals or numbers. All mammals past and future constitute an indefinitely large domain, and the numbers in our system of arithmetic are infinite.

In addition, our definition of a tautology would seem to be such as to preclude a restriction to a finite domain. A tautology is a proposition that is impossible to be false, and we usually understand by this an impossibility no matter what the size of the domain. A proposition that is a tautology in this general sense we shall refer to as an unrestricted tautology, in contrast to a restricted tautology that is necessarily true relative to a domain of a restricted number of individuals. Similarly, we can distinguish between an unrestrictedly valid inference, one valid no matter what the size of the domain, and a restrictedly valid inference, one that is valid only relative to a domain of a certain size. It is clear that it is the logical status of a general sentence and the validity of an inference in the unrestricted sense that we wish to be able to evaluate.

The truth table procedure just outlined does provide us, to be sure, with at least the theoretical means of evaluating sentences and inferences in this unrestricted sense. A general sentence will not express an unrestricted tautology if we can find a domain of a certain size relative to which the sentence can be interpreted as false. Nor will an inference be unrestrictedly valid if found invalid relative to a certain domain. It is often possible to find such a domain. Consider, for example, the inference,

$$\frac{\text{Some men are wise.}}{\therefore \text{ All men are wise.}}$$

For a domain of one individual this would be logically equivalent to an inference of the form

$$\frac{Pa \wedge Qa}{Pa \supset Qa}$$

and would be found to be valid relative to this domain in a truth table with four possible interpretations of the two constituents. With two individuals in the domain, however, its form would be

$$\frac{(Pa \wedge Qa) \vee (Pb \wedge Qb)}{(Pa \supset Qa) \wedge (Pb \supset Qb)}$$

When a truth table with 4 constituents and 16 possible interpretations is constructed it would be found to be invalid, e.g. when Pa, Qa, and Pb are true and Qb false. The inference is thus shown to be (unrestrictedly) invalid, for there is a domain of a certain size for which true premisses lead to a false conclusion.

To establish unrestricted invalidity (and also whether a sentence expresses an unrestricted contingent proposition) is thus possible by means of the standard truth table method of sentence logic. But notice how uncertain this method is. It depends on our being able to find a domain of the requisite size in which we have true premisses and a false conclusion. This may often be very difficult to accomplish. Considerable trial and error may be required to establish invalidity, unless we are for some reason fortunate enough to have guessed the size of the invalidating domain.

It can be even more difficult to show that an inference is unrestrictedly valid. Having shown that an inference is valid for a domain of a certain number n individuals, we have no general assurance that it will remain valid for a domain enlarged to n+1 individuals. Thus, an inference may be valid relative to a domain of four individuals, but conceivably be invalid if we add to the domain a fifth member. It has been proven, however, that if an inference has been shown to be valid for a domain of 2^m individuals, where m is the number of constituent predicates, then it will be unrestrictedly valid, no matter what the size of the domain, finite or infinite.[2] Thus, with two constituent monadic predicates an inference will be unrestrictedly valid provided it is valid relative to a domain of $2^2 = 4$ individuals. To establish this will require a truth table with 2 x 4 = 8 atomic constituents of which there are $2^8 = 256$ possible interpretations. It is evident that even for relatively simple inferences this decision procedure is extremely difficult to carry out. Though it is theoretically possible to decide unrestrictedly validity for any given inference, it is a practical impossibility for all but the simplest.

The difficulties with the procedure just outlined stem from its replacing universal and existential sentences with conjunctions or disjunctions of atomic sentences. This requires specifying particular individuals in the given domain to which the constituent predicates are ascribed. We turn now to two methods that overcome this difficulty by ignoring individual objects. The first is the decision procedure of the traditional Aristotelian logic, the second a decision procedure in which existential sentences of a certain form become the interpreted constituents.

Exercises. Evaluate the following inferences as valid or invalid relative to a domain of the size indicated in parentheses. Choose predicate and individual schemata that will facilitate the representation of logical form. For inference 1, for example, G may be chosen to represent '...is a gymnast', S to represent '...is strong', and j to represent the proper name 'Jack'.

1. All gymnasts are strong. Jack is not a gymnast. Therefore, he is not strong. (One individual)
2. All gymnasts are strong. Jack is not strong. Therefore, he is not a gymnast. (One individual)
3. Some men are Germans. Some men are wealthy. Therefore, some Germans are wealthy. (One individual)
4. Some trees have leaves. Some trees have branches. Therefore, some things with leaves have branches. (Two individuals)
5. All trees have leaves. Therefore, some trees have leaves. (One individual)
6. All trees have leaves. There are trees. Therefore, some trees have leaves. (One individual)
7. There are no unicorns. Therefore, all unicorns have horns. (Two individuals)
8. Everything is either round or red, since everything round is red. (Two individuals)
9. Nothing is heavy. Therefore, not everything is heavy. (Three individuals)

10. Something is not heavy. Therefore, not everything is heavy. (Three individuals)

21. Aristotelian Logic

Every sentence considered by the traditional Aristotelian logic is a general sentence with a subject term S joined to a predicate term P by the copula 'is'. Four forms of sentences were taken as fundamental in this logic and assigned the code letters A, E, I, and O. These are now listed along with their corresponding form in modern predicate logic.

    A (Universal Affirmative):   All S is P $\forall x(Sx \supset Px)$
    E (Universal Negative):   No S is P $\forall x(Sx \supset -Px)$
    I (Particular Affirmative): Some S is P $\exists x(Sx \wedge Px)$
    O (Particular Negative):   Some S is not P $\exists x(Sx \wedge -Px)$

Recall from Section 17 that we constructed a "Square of Opposition" between sentences of the form $\forall xPx$, $\forall x-Px$, $\exists xPx$, and $\exists x-Px$. A parallel diagram sums up the fundamental logical relationships between the four Aristotelian forms.

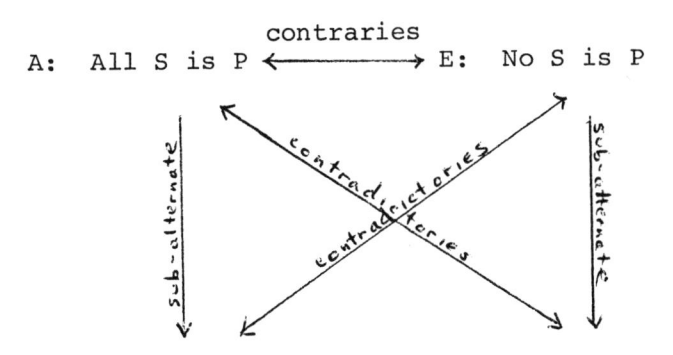

                       contraries
   A: All S is P ⟷ E: No S is P

(diagram: sub-alternate on sides, contradictories on diagonals)

   I: Some S is P ⟷ O: Some S is not P
              sub-contraries

As indicated in the diagram, the A and E forms are contraries, while I and O are sub-contraries. A and O are contradictories of one another, as are E and I. The relation of sub-alternation holds between the universal A and E forms and their corresponding E and O forms.

Immediate Inferences. The Aristotelian logic set as its task the evaluation of inferences whose premisses and conclusion could be represented by the four basic sentences. One class of inferences evaluated were the *immediate inferences*, those having but one premiss. Among these are inferences valid by "conversion," the operation of reversing the order of the subject and predicate terms. The following inference forms are valid by conversion.

$$\frac{\text{No S is P}}{\text{No P is S}} \quad \text{and} \quad \frac{\text{Some S is P}}{\text{Some P is S}}$$

Conversion is, however, impossible for the A and O forms. From All S is P it does not follow that All P is S, nor from Some S is not P does it follow that Some P is not S.

Another pair of valid immediate inferences are those between the universal affirmative and particular affirmative forms and between the universal negative and the particular negative, the relation of sub-alternation in the above Square of Opposition. The two inference forms are thus

$$\frac{\text{All S is P}}{\text{Some S is P}} \quad \text{and} \quad \frac{\text{No S is P}}{\text{Some S is not P}}$$

It was thought as "self-evident" that the universal A and E forms could only be true if there were Ss for them to be true of, and hence that the particular I and O forms could be inferred from them. The two inferences were thus valid because of what is called the "existential commitment" to objects that are S by the A and E forms.

This same commitment is not to be found in the corresponding forms in modern predicate logic, and sub-alternation thus becomes an invalid inference. The corresponding inference forms are

$$\frac{\forall x(Sx \supset Px)}{\exists x(Sx \wedge Px)} \quad \text{and} \quad \frac{\forall x(Sx \supset -Px)}{\exists x(Sx \wedge -Px)}$$

The definition of the material implication sign is such that $Sa \supset Pa$ is true if the antecedent is false, if a given individual a is not an S. From this it follows that $\forall x(Sx \supset Px)$ and $\forall x(Sx \supset -Px)$ will both be true if there are no Ss in the domain denoted by the variable, if for every individual it is not an S. The sentences 'All unicorns are white' and 'No unicorns are white', for example, will be true relative to the domain of material things, for no matter which of them we choose it will be false that it is a unicorn. But clearly if there are no Ss then the forms $\exists x(Sx \wedge Px)$ and $\exists x(Sx \wedge -Px)$ will be false. Since there are no Ss, there can be none that are both S and P nor none that are S and not P. Hence it is possible for sentences of the form $\forall x(Sx \supset Px)$ and $\forall x(Sx \supset -Px)$ to be true, while $\exists x(Sx \wedge Px)$ and $\exists x(Sx \wedge -Px)$ are false, and the sub-alternation forms are thus invalid.

The difference with regard to sub-alternation marks one of the major differences between Aristotelian and predicate logic. It can be seen to result from the innovation in predicate logic of paraphrasing universal sentences by first introducing a more general subject term to denote a domain and then converting the original subject to a predicate that is part of an antecedent of a material implication. No such paraphrase was regarded as essential in Aristotelian logic, the subject of a general sentence instead remaining the same after its form has been represented. Some logicians have thought sub-alternation to be an "intuitively valid" form of inference or one consistent with our "ordinary use of language," and have argued that its rejection marks a defect of modern logic.[1] But whatever conflict with our "logical intuitions" this rejection causes (and these intuitions can often prove misleading), this seems a small price to pay for the gain in generality that results when we adopt the paraphrases of predicate logic.

Syllogisms. The other class of inferences evaluated in the Aristotelian logic were the syllogisms, inferences with two premisses containing a third term called the "middle" term, that "mediates" the inference to the conclusion. The following is an example of an Aristotelian syllogism:

1) All men are mortal.
   All Greeks are men.
   ∴ All Greeks are mortal.

where 'men' is the middle term mediating between the subject of the conclusion 'Greeks' and the predicate 'mortal'. The form of 1) would be

2) All M is P
   All S is M
   All S is P

The subject of the conclusion S was called the "minor" term and its predicate P the "major" term. The first premiss contains the major term and was called the major premiss; the second premiss with the minor term was the minor premiss.

Dropping the quantifier 'all' in 2), it is clear that there are four possible arrangements of the middle term M and the minor and major terms S and P. These arrangements were called the four "figures." They are

```
   I.    M is P          II.    P is M
         S is M                 S is M
         ─────                  ─────
         S is P                 S is P

 III.    M is P          IV.    P is M
         M is S                 M is S
         ─────                  ─────
         S is P                 S is P
```

Each of the sentences occurring in the premisses and conclusion of the four figures can occur in one of the four possible A, E, I, and O forms. An inference of the form of 2) would be, for example, in the first figure with premisses and conclusion all of the A form. It is customary to refer to this as the AAA form of the first figure, or "AAA-I." With four possible sentence forms for the premisses and conclusion of each of the four possible figures it can be calculated that there are in all 256 possible syllogistic forms. The decision problem for the Aristotelian logic was that of distinguishing from within these 256 possible syllogistic forms those that are valid.

  Venn Diagrams. A solution to the problem was made possible by correlating general terms to classes of objects that comprise their extension and then diagramming relations that hold between these classes. To the general term 'man', for example, we can correlate the class of men that it denotes, and to 'mortal' we can correlate the class of mortal things. Then to say that all men are mortal is to say that the class of men is included in the class of those things that are mortal. Alternatively, it is to say that the class of men that are not mortal is empty or is identical with the null class, the class with no members. Let us represent 'men' by S and 'mortal' by P. The class of men that are not mortal (the intersection of the class of men with the class of things not mortal) we represent by $S\bar{P}$ and the null class by ϕ. Then to the sentence 'All S is P' we can correlate the identity $S\bar{P} = \phi$. The former will express a true proposition if and only if the identity between the class of Ss that are not Ps and the null class holds. By indicating that a class is empty by shading it this identity can be represented by the following "Venn diagram":

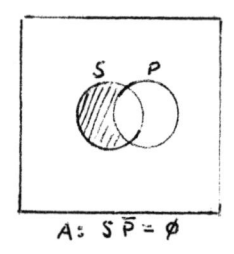

$$A: S\bar{P} = \phi$$

By indicating that the Ss not included in the class P is empty we have thus diagrammed the A form.

Similar diagrams are possible for the remaining three forms of the Square of Opposition. To say 'No S is P' is to assert that the class of Ss that are P is empty. Thus, 'No S is P' will be true if and only if SP = φ. 'Some S is P' will be true if and only if SP ≠ φ (with '≠' for 'is not identical with') or if the Ss that are P contains at least one member (is not empty). 'Some S is not P' will be true if and only if S$\bar{P}$ ≠ φ. With shaded areas indicating again the emptiness of a class and a cross indicating non-emptiness or the existence of at least one member in a class we have the following diagrams for the remaining forms:

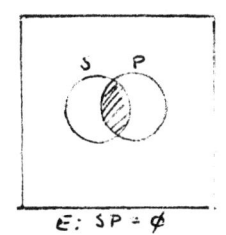

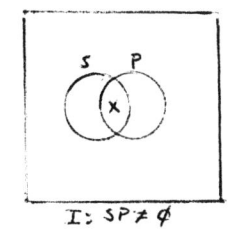

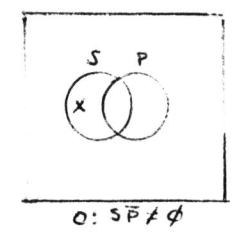

E: SP = φ   I: SP ≠ φ   O: S$\bar{P}$ ≠ φ

The procedure for evaluating a given syllogism is that of correlating to its premisses and conclusion their corresponding identities or non-identities between classes. These identities are then diagrammed in the manner just indicated in such a way as to show whether or not the premisses entail the conclusion. Inference 1) above would be evaluated, for example, by first diagramming the first premiss of the form All M is P and then superimposing on this diagram that for the second premiss All S is M. The successive diagrams would be

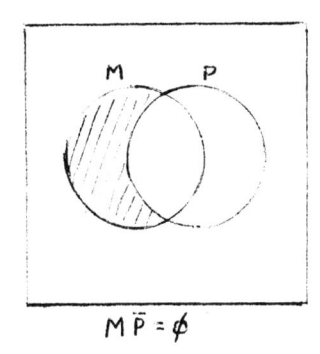

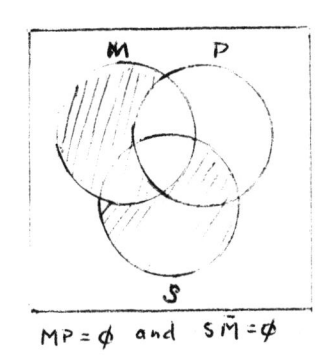

M$\bar{P}$ = φ   M$\bar{P}$ = φ and S$\bar{M}$ = φ

Since the right diagram shows that the area of S$\bar{P}$ is shaded, and hence S$\bar{P}$ = φ, it shows that the conclusion All S is P follows from the premisses. The syllogism AAA of the first figure (called the syllogism "Barbara") is thus shown to be valid.

In similar fashion we can evaluate an inference of the form

3) No P is M
   Some S is M
   ───────────
   No S is P

of the form EIE of the second figure. Its Venn diagram would be

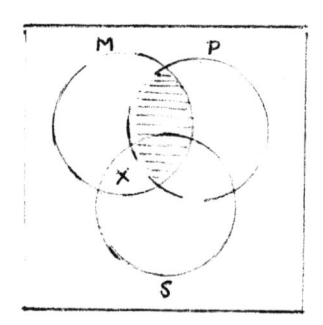

Since the intersection of S and P is not empty (SP ≠ φ) in the diagram, the conclusion does not follow, and the syllogism is shown to be invalid.

Note that this decision procedure does not require us to distinguish individuals within the extension of constituent terms, nor to restrict ourselves to classes of a certain definite size. The classes that are the extensions of the terms S, P, and M can have any number of members not excluding an infinite number. As a consequence of considering only relations between classes and ignoring individual members the Venn diagram procedure is considerably simpler than that of the preceding section. Note also that this procedure cannot validate the immediate inferences of sub-alternation, since shading out areas to indicate a class is empty is insufficient to indicate that a class contains at least one member. To validate the two sub-alternation inferences would require our adding the premiss that there exists an individual in the class denoted by the subject term.[2]

But this procedure is nevertheless severely restricted in applicability. It is designed for syllogisms containing three terms and two premisses and with constituent sentences in one of the four basic forms. It is possible to extend the Venn diagram method to inferences containing four terms and three premisses, but as we add the additional circle to our diagrams it becomes increasingly difficult to show whether or not the conclusion follows from the premisses by shading areas. It is also possible to "reduce" syllogisms with more than two premisses to standard forms with two premisses by a series of steps. But considerable ingenuity is usually required to effect such reductions, and they require departing from simple applications of the Venn diagram method. More serious still is the impossibility of evaluating within the Aristotelian logic inferences whose constituent sentences are not of the A, E, I, or O forms. On such inference would be

> Everything is both round and red.
> ∴ Everything is either round or red.

It is valid, though neither its premiss nor its conclusion is included in the four forms. To show its validity would require departing from the methods of Aristotelian logic.

For these reasons we return now to predicate logic and a decision procedure capable of evaluating an inference with any number of premisses and any number of constituent terms that are monadic predicates. The class of inferences that can be evaluated with this procedure is potentially infinite, not restricted to the 256 possible syllogistic forms. These moods can be regarded as a restricted sub-class of the possible inferences that can be evaluated within predicate logic.

Exercises. Evaluate the following syllogisms as either valid or invalid by means of Venn diagrams. Again choose subject and predicate symbols that will facilitate representation, e.g. M for 'mice' and T for 'with a tail'.

1. All mice have tails. Some rodents do not have tails. Therefore, some rodents are not mice.
2. All mice have tails. No rodents have tails. Therefore, no rodents are mice.
3. Some mice have tails. Some rodents do not have tails. Therefore, some mice are not rodents.
4. All mice have tails. All mice are rodents. Some rodents have tails.
5. All rodents are animals. All mice are rodents. Therefore, some mice are animals.
6. No reptiles can fly. All snakes are reptiles. Therefore, no snakes can fly.
7. No reptiles can fly. Some snakes are reptiles. Therefore, some snakes cannot fly.
8. No reptiles can fly. Some bats can fly. Therefore, some bats are not reptiles.

22. Distributive Normal Forms*

A decision procedure requires our being able to give an exhaustive enumeration of the possible interpretations of the constituent elements of the inference. To determine the validity of an inference or the logical status of a sentence requires our being able to survey these possibilities. For sentence logic these constituent elements are the recurring atomic sentences, and the enumeration of possibilities is in a truth table. In Section 20 we outlined a method of extending this procedure to predicate logic in which quantificational schemata were expressed as truth functions of atomic schemata. In the present section we develop a decision procedure based on possible interpretations of constituents that are themselves quantificational schemata.[1]

This different system of interpretation can be illustrated by a simple example. Suppose we have a single term, say 'red', and wish to apply it to a domain of objects. Then there would seem to be two existential sentences that would apply to the domain: 'Something (at least one) is red' and 'Something is not red'. And there would seem to be these four possibilities:

  i) It is true that something is red and true that something is not red.
 ii) It is true that something is red, while false that something is not red.
iii) It is false that something is red and true that something is not.
 iv) It is false both that something is red and something is not.

Possibility i) would be realized if some of the objects in the domain were red, while others green or some other color; ii) would hold if all the objects were red; iii) would hold if none of the objects were red; while iv) would hold if there were no objects in the domain to which a color word like 'red' is applicable. Representing the predicate '...is red' by P and using our symbolism for existential quantification, the four possibilities may be represented in a truth table.

---

*This section along with Sections 23 and 24 may be skipped on a first reading in order that the reader may turn directly to Section 25 and the predicate calculus in Chapter VI. The decision procedure of these three sections is much more difficult to apply than the predicate calculus. Pp. 110-112 in Section 24 should be returned to, however, prior to reading the application of the predicate calculus to relations in Section 28.

$\exists xPx$	$\exists x-Px$
T	T
T	F
F	T
F	F

These are exclusive and exhaustive possibilities: no two of them can both hold for a given domain and at least one of them must hold, no matter what the particular characteristics of the objects of the domain may happen to be. Notice that the size of the domain is irrelevant to the enumeration of these possibilities; they hold for a domain of any number of members, finite or infinite. The fourth possibility requiring an empty domain is usually disregarded in logic, the assumption being made that the domain is non-empty and that there is hence at least one object for which either a given predicate or its negation holds (cf. end of Section 17).

Now suppose we had two terms, say 'red' and 'green', and wish to apply both to our domain of objects. Then there would clearly be four existential sentences that would apply to the domain: 'Something is both red and green', 'Something is red and not green', 'Something is not red and green', and 'Something is not red and not green'. In each of these sentences occur each of the two terms or its negation. Representing the predicate '...is red' again by P and '...is green' by Q we should have now altogether sixteen possible interpretations which again can be put in tabular form.

$\exists x(Px \land Qx)$	$\exists x(Px \land -Qx)$	$\exists x(-Px \land Qx)$	$\exists x(-Px \land -Qx)$
T	T	T	T
T	T	T	F
T	T	F	T
T	T	F	F
T	F	T	T
T	F	T	F
T	F	F	T
T	F	F	F
F	T	T	T
F	T	T	F
F	T	F	T
F	T	F	F
F	F	T	T
F	F	T	F
F	F	F	T
F	F	F	F

Once again we can exclude the possibility of an empty domain and the last possibility of all the existential schemata being false. The fifteen remaining possibilities are again exclusive and exhaustive.

Note that for one predicate schema P there are two existential schemata that can be formed and three possible interpretations of them if the domain is assumed to be non-empty. For two predicates P and Q there are four existential sentences and fifteen possible interpretations. In general, n predicates will generate 2^n existential sentences of which there are $2^{2^n}-1$ exclusive and exhaustive possible interpretations if a non-empty domain is assumed. Within these 2^n existential sentences will occur each of the n predicates or its negation. We shall call the existential schemata representing these sentences <u>existence-constituents</u> or $\exists$-<u>constituents</u>.

In sentence logic we could enumerate the possible interpretations of atomic sentences as the basic constituents of inferences and molecular sentences. Such an enumeration made possible a decision procedure. We can likewise enumerate, as we have seen, the possible interpretations of our $\exists$-constituents. We can also transform every quantificational schema containing n monadic predicates into a logically equivalent schema that is a truth function of one or more of the 2^n $\exists$-constituents formed from the n predicates. These logically equivalent schemata are called the quantificational schema's <u>distributive normal form</u>. The fact that every quantificational schema can be transformed into a distributive normal form makes possible a truth table decision procedure for predicate logic when applied to monadic predicates.

The transformation into distributive normal form is made possible by three logical facts: i) every universal quantificational schema can be expressed as an existential schema; ii) every open schema is logically equivalent to a disjunction or conjunctions of each of its n constituents or their negations, or can be transformed into its disjunctive normal form; and iii) the existential quantifier can be distributed over the disjunction sign. The expansion of a schema into its distributive normal form is effected in three stages by means of these logical facts. We consider these stages in order.

<u>Universal to Existential Schemata</u>. First let us adopt a notation that will make our transformations easier. Let us rewrite quantificational schemata in a form in which variables are dropped, conjunction between open schemata within the scope of a quantifier indicated by juxtaposition, and the negation sign operating on an open schema rewritten as a bar over the negated schema. We shall retain our standard negation and conjunction signs for operating on or combining quantificational schemata. With these abbreviations $\forall x(Px \supset Qx)$ becomes rewritten as $\forall(P \supset Q)$; $\exists x(Px \land -Qx)$ becomes $\exists(P\overline{Q})$; $-\exists x(Px \land -Qx) \land \forall xQx$ becomes $-\exists(P\overline{Q}) \land \forall Q$.

To rewrite universal schemata as existential schemata we simply appeal to the logical equivalence between $\forall x Px$ and $-\exists x -Px$ noted at the end of Section 17. Thus, $\forall x(Px \supset Qx)$ when rewritten as $\forall(P \supset Q)$ is equivalent to and can be replaced by $-\exists(P \supset Q)$.

<u>Disjunctive Normal Forms of Open Schemata</u>. In Section 10 we developed a procedure for expanding sentence schemata into a so-called "disjunctive normal form" as a disjunction of a certain kind of conjunctions. The same procedure can be extended to open schemata, and the results used in the expansion of a quantificational schema into its distributive normal form.

The expansion procedure is enabled by the fact that to every predicate we can correlate the class of objects which constitutes its extension. For example, we can correlate to the predicate '...is blue' the class of all blue objects. In terms of this correspondence between predicates and classes we are able to define the relation of logical equivalence between two predicates: two predicates P and Q can be said to be logically equivalent if they have the same extension, or $P \Leftrightarrow Q$ if and only if P and Q have the same extension.

It is evident from this definition of logical equivalence that we can state that $\bar{\bar{P}} \Leftrightarrow P$, since the extension of the double negation of a predicate is the same as that for the predicate itself (the class of all objects for which it is not the case that they are not blue is the class of all blue objects). It can, in fact, be easily shown that all the basic equivalences between sentence schemata listed in Section 10 also hold for predicate schemata, with the tautologous proposition ◻ being interpreted as the predicate holding of all objects in a given domain and the contradictory proposition φ the predicate whose extension is no objects, the null class. Thus $PP \Leftrightarrow P$, $P \vee P \Leftrightarrow P$ (idempotent laws), $PQ \Leftrightarrow QP$, $P \vee Q \Leftrightarrow Q \vee P$ (commutative laws), etc. That these equivalences do hold can be shown by diagramming the classes that constitute the extension of the predicates and showing that they coincide. The equivalence $\overline{(P \vee Q)} \Leftrightarrow \overline{P}\overline{Q}$ is established, for example, by the following diagram:

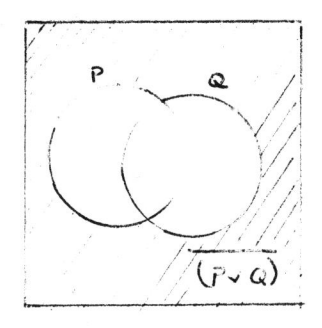

Here it is seen that the class of objects that P or Q does not hold of, $\overline{(P \vee Q)}$, is identical with those objects that are neither P nor Q, $\overline{P}\overline{Q}$. Since the extension of the two predicates is the same, they are shown to be logically equivalent.

Adopting also an analogue to the rule of replacement of sentence logic (cf. Section 10) allows us to expand any given predicate schema with n constituents into its disjunctive normal form as a disjunction of conjunctions containing each of the n constituent predicate schemata or its negation. For example, the steps in the expansion of the predicate schema P ∨ Q are

$$P \vee Q$$
$$P(Q \vee \overline{Q}) \vee Q(P \vee \overline{P})$$
$$PQ \vee P\overline{Q} \vee PQ \vee \overline{P}Q$$
$$P\overline{Q} \vee PQ \vee \overline{P}Q$$

The procedural steps outlined in Section 10 where A ∨ B is expanded will apply here, with predicate schemata replacing throughout sentence schemata. Each step is derived from the preceding by replacing predicate schemata by their logical equivalents.

We are now in a position to define logical equivalence between open schemata: two open sentences represented by Px and Qx can be said to be logically equivalent if their constituent predicates P and Q are equivalent, or Px ⇔ Qx if and only if P ⇔ Q. By means of this correlation between equivalence between open schemata and equivalence between predicate schemata we can directly extend our expansion of predicate schemata into their disjunctive normal forms to their corresponding open schemata. Thus, if the disjunctive normal form of the predicate schema P ∨ Q is $PQ \vee P\overline{Q} \vee \overline{P}Q$, the logically equivalent disjunctive normal form of the open schema Px ∨ Qx will be (Px ∧ Qx) ∨ (Px ∧ -Qx) ∨ (-Px ∧ Qx). Since every open schema has a logically equivalent disjunctive normal form, we can replace an open schema within the scope of a quantifier by its disjunctive normal form. For example, since Px ⊃ Qx abbreviated as P ⊃ Q is equivalent to $PQ \vee \overline{P}Q \vee \overline{P}\overline{Q}$ (also abbreviated by dropping variables, juxtaposition for conjunction, and negation as a bar), ∃(P ⊃ Q) is logically equivalent to ∃($PQ \vee \overline{P}Q \vee \overline{P}\overline{Q}$). Our being able to replace open schemata within the scope of a quantifier by their disjunctive normal form enables the final stage in the expansion of a quantificational schema into its distributive normal form.

<u>Distribution of the Existential Quantifier</u>. This final step is made possible by the fact that the existential quantifier distributes over the disjunction sign. For it is evident that if a sentence such as 'Something is red or green' is true, then so is 'Something is red or something is green'. More generally, if any existential sentence of the form ∃x(Px ∨ Qx ∨ ... ∨ Zx) is true,

then the disjunction of existential sentences $\exists xPx \lor \exists xQx \lor \ldots \lor \exists xZx$ is also true.

To illustrate the expansion procedure we take the schema

    1) $\forall x(Px \land Qx)$

Using our abbreviating conventions 1) can be rewritten as

    2) $\forall(PQ)$

Rewriting this universal schema as an existential schema we get

    3) $-\exists \overline{(PQ)}$

Now we must transform the open schema of 3), $\overline{(PQ)}$, into its disjunctive normal form. It happens to be (as can easily be verified by the reader) $\overline{P}Q \lor P\overline{Q} \lor \overline{P}\overline{Q}$. Replacing this for the schema $\overline{(PQ)}$ in 3) we derive

    4) $-\exists(\overline{P}Q \lor P\overline{Q} \lor \overline{P}\overline{Q})$

By distributing the existential quantifier over the disjuncts of 4) we finally obtain

    5) $-[\exists(\overline{P}Q) \lor \exists(P\overline{Q}) \lor \exists(\overline{P}\overline{Q})]$

This expression is now a truth function of the $\exists$-constituents $\exists(\overline{P}Q)$, $\exists(P\overline{Q})$, and $\exists(\overline{P}\overline{Q})$, and is hence in distributive normal form.

In general, every quantificational schema can be transformed into a logically equivalent distributive normal form in a similar fashion. The steps in the transformation are: i) abbreviate the quantificational schema by dropping variables and adopting new conventions for negation and conjunction; ii) rewrite universal schemata as existential schemata; iii) replace open schemata of the existential schemata by their disjunctive normal forms; and iv) distribute the existential quantifier over the resulting disjunctions of conjunctions.

Exercises. Expand the following schemata into their distributive normal forms.

1. $\forall x(Px \lor Qx)$
2. $\forall x(Px \supset Qx)$
3. $\exists x(Px \lor Qx)$
4. $\forall x(Px \equiv Qx)$
5. $\exists x(Px \supset Qx) \lor \forall xPx$
6. $\forall x(Px \land Qx \supset Rx)$
7. $\forall x(Px \supset -Qx) \land \exists x(Qx \land Rx)$
8. $-\forall x(Px \land -Qx) \lor \exists x(Qx \lor -Rx)$
9. $\exists x(Px \land Qx) \supset \exists x(Qx \lor Rx)$
10. $\exists x(-Px \lor Qx) \equiv \forall x(Px \lor Qx \supset Rx)$

23. Decision Procedure for Predicate Logic

With the means at hand of transforming a quantificational schema into its distributive normal form, a decision procedure can be readily constructed in which $\exists$-constituents serve as the constituents for which we list possible interpretations. The validity or invalidity of an inference and more generally whether or not a general sentence expresses a tautology[1] can be determined relative to these possible interpretations. It must be emphasized again, however, that this procedure will be considered in this section only relative to inferences and general sentences whose constituents are monadic predicates. As we shall see in the next section, difficulties confront us when we attempt to extend it to relational predicates.

Evaluation of Inferences. To see how we can evaluate an inference, let us take as our example the valid inference,

> Everything is red and green.
> ∴ Something is red or green.

of the form,

$$\frac{\forall x(Px \wedge Qx)}{\exists x(Px \vee Qx)}$$

To evaluate this inference we must first expand the premiss and conclusion into their distributive normal form. The premiss when abbreviated becomes

1) $\forall(PQ)$

Its distributive normal form, as we saw in the previous section, is

2) $-[\exists(P\overline{Q}) \vee \exists(\overline{P}Q) \vee \exists(\overline{P}\overline{Q})]$

The conclusion when abbreviated is

3) $\exists(P \vee Q)$

To expand 3) into its distributive normal form we first replace its open schema $P \vee Q$ by its disjunctive normal form $PQ \vee P\overline{Q} \vee \overline{P}Q$ to obtain

4) $\exists(PQ \vee P\overline{Q} \vee \overline{P}Q)$

We then distribute the existential quantifier over the disjuncts within 4) to obtain the distributive normal form

5) $\exists(PQ) \vee \exists(P\overline{Q}) \vee \exists(\overline{P}Q)$

With both premiss and conclusion in the distributive normal forms obtained in 2) and 5) we can construct a truth table enumerating the possible interpretations of the four $\exists$-constituents generated by the two predicates.

$\exists(PQ)$	$\exists(P\overline{Q})$	$\exists(\overline{P}Q)$	$\exists(\overline{P}\overline{Q})$	$\sim\{\exists(P\overline{Q})\vee[\exists(\overline{P}Q)\vee\exists(\overline{P}\overline{Q})]\}$			$\exists(PQ)\vee[\exists(P\overline{Q})\vee\exists(\overline{P}Q)]$	
T	T	T	T	F	T	T	T	T
T	T	T	F	F	T	T	T	T
T	T	F	T	F	T	T	T	T
T	T	F	F	F	T	F	T	T
T	F	T	T	F	T	T	T	T
T	F	T	F	F	T	T	T	T
T	F	F	T	F	T	T	T	F
T	F	F	F	T	F	F	T	F
F	T	T	T	F	T	T	T	T
F	T	T	F	F	T	T	T	T
F	T	F	T	F	T	T	T	T
F	T	F	F	F	T	F	T	T
F	F	T	T	F	T	T	T	T
F	F	T	F	F	T	T	T	T
F	F	F	T	F	T	T	F	F
~~F~~	~~F~~	~~F~~	~~F~~	~~T~~	~~F~~	~~F~~	~~F~~	~~F~~

If we assume a non-empty domain and exclude the last interpretation (this is indicated by drawing a line through this interpretation), there is no interpretation under which the premiss is true and the conclusion fasle. The premiss thus entails the conclusion, and the inference is shown to be valid.

In the above truth table there occur all the possible 2^n $\exists$-constituents that are generated by n predicates and all of the $2^{2^n}-1$ possible interpretations. For certain inferences it is possible to shorten these truth tables. Consider, for example, the Aristotelian syllogism 'All men are mortal. All Greeks are men. ∴ All Greeks are mortal' of the form

6) $\quad\dfrac{\begin{array}{l}\forall x(Qx\supset Rx)\\ \forall x(Px\supset Qx)\end{array}}{\forall x(Px\supset Rx)}$

To evaluate it we would transform the premisses and conclusion of 6) into their distributive normal forms. Since the constituent predicate schemata of 6) are P, Q, and R, $\exists$-constituents must contain each of these three predicates or their negations. The expansion of the first premiss is

7) $\forall (Q \supset R)$ (abbreviated symbolism)
   $-\exists \overline{(Q \supset R)}$ (rewriting the universal schema as an existential schema)
   $-\exists (PQ\overline{R} \vee \overline{P}Q\overline{R})$ (replacing the open schema by its disjunctive normal form)
   $-[\exists (PQ\overline{R}) \vee \exists (\overline{P}Q\overline{R})]$ (distributing existential quantifier over disjuncts)

The expansion of the second premiss, $\forall x(Px \supset Qx)$, is similar.

8) $\forall (P \supset Q)$
   $-\exists \overline{(P \supset Q)}$
   $-\exists (P\overline{Q}R \vee P\overline{Q}\overline{R})$
   $-[\exists (P\overline{Q}R) \vee \exists (P\overline{Q}\overline{R})]$

The conclusion is also expanded in the same way.

9) $\forall (P \supset Q)$
   $-\exists \overline{(P \supset Q)}$
   $-\exists (P\overline{Q}R \vee P\overline{Q}\overline{R})$
   $-[\exists (P\overline{Q}R) \vee \exists (P\overline{Q}\overline{R})]$

Notice that the distributive normal forms obtained in 7), 8), and 9) contain only four of the $2^3=8$ possible $\exists$-constituents generated by three predicates. It is possible in such a case to omit from our truth table the $\exists$-constituents not found in the distributive normal forms, and hence to reduce the number of possible interpretations. If we label the $\exists$-constituents heading the columns in the truth table by ①, ②, ③, and ④ and abbreviate premisses and conclusion as a truth function of these representing numbers, the truth table would have the following appearance:

① $\exists(PQ\overline{R})$	② $\exists(\overline{P}Q\overline{R})$	③ $\exists(P\overline{Q}R)$	④ $\exists(P\overline{Q}\overline{R})$	$-(① \vee ②)$	$-(③ \vee ④)$	$-(① \vee ④)$
T	T	T	T	F T	F T	F T
T	T	T	F	F T	F T	F T
T	T	F	T	F T	F T	F T
T	T	F	F	F T	T F	F T
T	F	T	T	F T	F T	F T
T	F	T	F	F T	F T	F T
T	F	F	T	F T	F T	F T
T	F	F	F	F T	T F	F T
F	T	T	T	F T	F T	F T
F	T	T	F	F T	F T	T F
F	T	F	T	F T	F T	F T
F	T	F	F	F T	T F	T F
F	F	T	T	T F	F T	F T
F	F	T	F	T F	F T	T F
F	F	F	T	T F	F T	F T
F	F	F	F	T F	T F	T F

For none of the possible interpretations of the $\exists$-constituents are the premisses true and the conclusion false, and the syllogism is thus shown to be valid. Note that in the last row of this truth table all the $\exists$-constituents are interpreted as false. With less than 2^n columns of $\exists$-constituents this interpretation does not imply an empty domain, and is therefore not excluded as a possible interpretation.

General Decision Procedure. As before for sentence logic it is possible to extend our decision procedure beyond inferences to the evaluation of the logical status of any general sentence. In particular, our procedure can decide whether the sentence,

10) Everything is both red and green if and only if everything is red and everything is green

of the form

11) $\forall x(Px \wedge Qx) \equiv \forall xPx \wedge \forall xQx$

expresses a tautology or not. The distributive normal form of 11) proves to be (as can be verified by the reader)

$-[\exists(\overline{PQ}) \vee \exists(P\overline{Q}) \vee \exists(\overline{P}\overline{Q})] \equiv -[\exists(\overline{P}Q) \vee \exists(\overline{P}\overline{Q})] \wedge -[\exists(P\overline{Q}) \vee \exists(\overline{P}\overline{Q})]$

When a truth table with 8 possible interpretations of the three $\exists$-constituents in the above expression is constructed we find that the expression is true for all of them. 10) thus expresses an (unrestricted tautology), and hence $\forall xPx \wedge \forall xQx$ and $\forall x(Px \wedge Qx)$ are shown to represent sentences expressing logically equivalent propositions.

Where the disjunctive normal form within the scope of an existential quantifier is a contradiction, we must assume that the existential schema is necessarily false. For example, $\forall x(Px \vee -Px)$ in abbreviated form is $\forall(P \vee -P)$. This is equivalent to $-\exists\overline{(P \vee P)}$, which in turn is equivalent to $-\exists(P\overline{P})$. Since $P\overline{P}$ is a contradiction, we know that $\exists(P\overline{P})$ is a contradiction, and hence that $-\exists(P\overline{P})$ is a tautology.

Using these procedures we can decide for any general sentence containing n monadic predicates whether it expresses an unrestricted tautology, a contingent proposition, or a contradiction. If it is true for all $2^{2^n}-1$ interpretations of its 2^n possible $\exists$-constituents, the sentence expresses a tautology; if it is true for some of these interpretations and false for others, it is contingent; if it is false for all, it expresses a contradiction. Among the tautologies that can be established by this procedure are those stating the distribution of the quantifiers over conjunction and disjunction. The fact that a sentence of the form of 11) is a tautology shows that the universal quantifier can be distributed over the conjunction sign. Besides 11) we have the following as tautologies:

$$\forall xPx \lor \forall xQx \supset \forall x(Px \lor Qx)$$
$$\exists x(Px \land Qx) \supset \exists xPx \land \exists xQx$$

As can be verified, $\forall x(Px \lor Qx)$ does not entail $\forall xPx \lor \forall xQx$, nor does $\exists xPx \land \exists xQx$ entail $\exists x(Px \land Qx)$. Logical equivalence holds only for the distribution of the universal quantifier over conjunction and the existential quantifier over disjunction.

Predicates and Sentences as Recurring Constituents. We can easily extend the decision procedure just outlined to inferences and sentences whose validity and logical status depends upon the recurrence of both constituent (monadic) predicates and sentences. Consider, for example, the inference,

12) If Jones is telling the truth, someone in the hotel committed the crime.
Jones is not telling the truth.
∴ Nobody was in the hotel.

The recurring constituents can be seen to include the whole sentence 'Jones is telling the truth' in addition to the predicates '...was in the hotel' and '...committed the crime'. Its logical form is

13) $A \supset \exists x(Px \land Qx)$
$-A$
$\overline{\forall x-Px}$

with the sentence schema A for 'Jones is telling the truth'. Written with its premisses and conclusion abbreviated and put in distributive normal form, 13) becomes

$A \supset \exists(PQ)$
$-A$
$\overline{-[\exists(PQ) \lor \exists(P\overline{Q})]}$

A truth table can now be constructed in which the possible interpretations of the sentence schema A and the two $\exists$-constituents are listed.

A	① $\exists(PQ)$	② $\exists(P\overline{Q})$	$A \supset ①$	$-A$	$-(① \lor ②)$
T	T	T	T	F	F	T
T	T	F	T	F	F	T
T	F	T	F	F	F	T
T	F	F	F	F	T	F
F	T	T	T	T	F	T
F	T	F	T	T	F	T
F	F	T	T	T	F	T
F	F	F	T	T	T	F

The inference is shown to be invalid by the fifth, sixth, and seventh interpretations.

Exercises.

I. Evaluate the following inferences by means of distributive normal forms. In 8-10 both sentences and predicates are the recurring constituents. Notice that the representation of some of the inferences does not require introducing category terms as subjects.

1. All children are small. Therefore, some children are small.
2. No children are small. There is at least one child. Therefore, some child is not small.
3. All children are small. Therefore, only children are small.
4. Everything has mass if and only if it is extended. There exist things with mass. Therefore, there is nothing extended that lacks mass.
5. There are no unicorns. Therefore, all unicorns have horns.
6. Some voters are not intelligent. This is because only voters are adults and not all adults are intelligent.
7. All adults who vote are responsible citizens. Therefore, if anyone is an adult, then if he is not a responsible citizen he does not vote.
8. If someone studies, the test will be an easy one. Therefore, no one will study, since the test will not be an easy one.
9. If someone studies he will pass the test. If no one studies but passes the test, the test will be an easy one. The test will not be easy. Therefore, no one will pass the test.
10. If Descartes is right, then nothing constitutes knowledge unless it is certain. If everything is certain then there is no use for the word "certain." Therefore, if there is a use for "certain," Descartes is mistaken.

II. Determine whether or not the following schemata represent sentences expressing unrestricted tautologies, contradictions, or contingent propositions.

1. $-\forall x(Px \supset Px)$
2. $\forall x(Px \supset -Px)$
3. $\forall x(Px \supset Qx) \supset (\forall xPx \supset \forall xQx)$
4. $(\forall xPx \supset \forall xQx) \supset \forall x(Px \supset Qx)$
5. $\forall x(Px \supset Qx) \supset (\exists xPx \supset \exists xQx)$
6. $(\exists xPx \supset \exists xQx) \land -\exists x(Px \supset Qx)$
7. $\exists x(Px \supset Qx) \equiv \exists xPx \supset \exists xQx$
8. $\exists x(Px \supset Qx) \equiv \forall xPx \supset \exists xQx$
9. $(\exists xPx \land \forall xQx) \land -\exists x(Px \land Qx)$
10. $-\forall x(Px \land Qx \supset Rx) \land -\exists x(Qx \land Rx)$
11. $\forall x(Px \lor Qx \supset Rx) \supset -\exists x(Qx \land -Rx)$
12. $\forall x(Px \land Qx \supset Rx) \equiv \forall x[Px \supset (Qx \supset Rx)]$

24. Relational Predicates

It has been stressed that the decision procedures outlined so far in this chapter have been restricted to inferences and sentences whose constituents are monadic or one-place predicates. We have not yet attempted to evaluate an inference such as

1) John is married to Alice.
∴ Alice is married to John.

whose recurring constituents include the relational two-place or dyadic predicate '...is married to...' expressing a relation holding between individuals. Can the procedures of Sections 20-23 applicable to monadic predicates be extended in order to evaluate inferences such as 1)? The answer turns out to be a negative one. In contrast to monadic predicates, relational predicates are peculiarly resistant to our attempts at evaluation.

The reason for our inability to evaluate an inference such as 1) by means of the Venn diagrams of Aristotelian logic should be obvious. The Aristotelian logic considered only inferences whose constituent premisses and conclusion are sentences with a single subject, and the Venn diagrams represent only relations between classes of objects that share a common attribute but are otherwise unrelated. The reason for the breakdown of the procedures of modern predicate logic is more difficult to determine. To see why they cannot be applied we must first observe that relational predicates can be classified into different types in a way not found for monadic predicates. These types will be determined by the kind of relation the predicate expresses.

Types of Dyadic Predicates. Consider the dyadic predicate '...is the cousin of...'. This expresses a relation which if holding between two individuals a and b also holds between b and a. If John is the cousin of Bill then it follows that Bill is the cousin of John. Such a relation is said to be a _symmetric relation_. More generally, for any dyadic predicate R, R is said to express a symmetric relation under the condition that $\forall x \forall y (Rxy \supset Ryx)$. Other predicates expressing symmetric relations would be '...is next to...', '...is the sibling of...', etc. A relation is said to be _asymmetric_ if when holding between a and b it does not hold between b and a, or more generally if $\forall x \forall y (Rxy \supset -Ryx)$. '...is the father of...' expresses an asymmetric relation, for if John is the father of Bill then Bill is not John's father. Other asymmetric relations are those of being greater than, taller than, uncle of, etc. A relation is said to be _nonsymmetric_ if it is neither symmetric nor asymmetric, e.g. the relation expressed by '...loves...', since John's loving Mary does not insure either her returning the favor or withholding it. One way of classifying dyadic predicates is thus in terms of the symmetry of the relations they express.

Another way is in terms of the transitivity of the expressed relations. A relation expressed by a dyadic predicate R is <u>transitive</u> if when holding between individuals a and b and b and c it also holds between a and c, or if $\forall x \forall y \forall z (Rxy \land Ryz \supset Rxz)$. '...is larger than...' would be an example of such a predicate, for if a is larger than b and b is larger than c, a is larger than c. Other examples would be '...is north of...' and '...is later than...'. A dyadic predicate R expresses an <u>intransitive</u> relation when $\forall x \forall y \forall z (Rxy \land Ryz \supset -Rxz)$, as for '...is the father of...'. Relations neither transitive nor intransitive are <u>nontransitive</u>.

A third classification of dyadic predicates is with respect to reflexivity. A <u>reflexive</u> relation is one which if the relation holds between any two things it holds between anything and itself. Thus, a relation is reflexive if and only if $\forall x \forall y (Rxy \supset Rxx \land Ryy)$. A relation that every individual bears to itself or for which $\forall x Rxx$ is said to be a <u>totally reflexive</u>, e.g. that expressed by '...is identical with...'. It is evident that all totally reflexive relations are reflexive but that not all reflexive relations are totally reflexive. Being in the same family is not totally reflexive, since not everything, e.g. rocks, is in the same family as itself. It is, however, reflexive, for if this relation does hold between two things, e.g. between two humans, it holds between the things and themselves. Reflexive relations can be made into totally reflexive relations only by restricting the domain to those things between which the relation holds, e.g. from material things in general to organisms. An <u>irreflexive</u> relation is one for which $\forall x -Rxx$, e.g. that expressed by '...is greater than...' or '...is the father...'. A <u>nonreflexive</u> relation is one which is neither reflexive nor irreflexive.

Invariably a given dyadic predicate can be classified in all three of the ways just described. The predicate '...is greater than...', for example, expresses an asymmetric, transitive, and irreflexive relation, as does '...is the successor of...' applied to the integers. Such predicates as '...is the same color as...' and '...is divisible by the same number as...' express relations that are symmetric, transitive, and reflexive, or what are referred to as "equivalence relations."

Besides classifying relational predicates with respect to the symmetry, transitivity, and reflexivity of the relations they express, we can also trace certain relations between these predicates. Perhaps the most important is that between dyadic predicates expressing relations that are said to be the <u>converse</u> of one another. The relation expressed by predicates R and S are said to be the converse of one another if R holds between two individuals a and b if and only if S holds between b and a, or if $\forall x \forall y (Rxy \equiv Syx)$. '...is the teacher of...' is thus the converse of '...is the student of...', since a is the teacher of b if and only if b is the student of a. Other relations that are converses of one another are those expressed by '...wins over...' and '...loses to...' and '...is greater than...' and '...is less than...'. By transforming a transitive verb into its passive form we can derive predicates bearing this relation to one another, e.g. by deriving from '...hits...' '...is hit by...', from '...loves...', '...is loved by...', etc. A symmetric relation may be defined as a relation that is its own converse.

Finite Domains. The fact that there are differences and relations between relational predicates that depend on their meanings, on the kind of relation that they express, makes it impossible to evaluate many inferences in which they occur as valid or invalid by virtue of their form alone. When we restrict our domain to a finite domain, however, a decision procedure of the kind outlined for monadic predicates in Section 20 is available to us, since we can again express universal and existential sentences as conjunctions and disjunctions of atomic sentences. Suppose, for example, that our domain consists of two individuals a and b and we have a universal sentence of the form $\forall x \forall y Rxy$. The conjunction to which this is equivalent would be one for which all the possible relations between the two individuals are included, that is,

2) $\quad \forall x \forall y Rxy \Leftrightarrow Rab \wedge Rba \wedge Raa \wedge Rbb$

The equivalence can be seen to state that $\forall x \forall y Rxy$ is true if and only if the relation expressed by R is symmetric and totally reflexive. Similarly, the existential schema $\exists x \exists y Rxy$ is logically equivalent to $Rab \vee Rba \vee Raa \vee Rbb$ for a domain of two individuals. $\exists x \exists y Rxy$ is thus false if and only if R is neither symmetric nor totally reflexive.

These equivalences can be used in the evaluation of an inference of the form

3) $\quad \dfrac{\forall x \forall y (Rxy \supset -Ryx)}{\forall x \forall y -Rxx}$

where we assume a domain of two individuals. Replacing premiss and conclusion by logically equivalent conjunctions of atomic schemata, 3) becomes

4) $\quad \dfrac{(Rab \supset -Rba) \wedge (Rba \supset -Rab) \wedge (Raa \supset -Raa) \wedge (Rbb \supset -Rbb)}{-Raa \wedge -Rbb}$

A truth table with columns headed by the atomic constituents Rab, Rba, Raa, and Rbb would show 3) to be a valid inference form, since Raa -Raa entails -Raa and Rbb -Rbb entails -Rbb. The evaluation of 3) thus establishes that if a relation is asymmetric (relative to a domain of two individuals) it must necessarily be irreflexive.

The equivalence between universal and existential relational sentences and conjunctions and disjunctions of atomic sentences can also be used to evaluate an inference such as 1) above. Representing '...is married to...' by R, the form of 1) is

$$\frac{Rab}{Rba}$$

A direct application of the truth table procedure would, of course, show this valid inference to be invalid. But by specifying in an additional premiss that '...is married to...' expresses a symmetric relation we can evaluate 1) as valid. Introducing as a premiss the definition of a symmetric relation, the form of 1) now becomes

5) $$\frac{\forall x \forall y (Rxy \supset Ryx)}{\dfrac{Rab}{Rba}}$$

Since there are only two individuals specified in the inference, we can restrict our domain to two individuals. Replacing its first premiss with its logically equivalent conjunction 5) becomes

$$\frac{(Rab \supset Rba) \land (Rba \supset Rab) \land (Raa \supset Raa) \land (Rbb \supset Rbb)}{\dfrac{Rab}{Rba}}$$

A truth table will now show this to be valid, since the conjunction entails the conjunct $Rab \supset Rba$ which together with Rab entails Rba.

By regarding them as enthymemes (cf. Section 8) any inference such as 1) with singular premisses expressing relations can be evaluated in this manner. We first add as the missing implied premiss one specifying the type of relation expressed within the premiss. We then regard this premiss as restricted to a domain with the same number as the singular terms occurring within the inference and replace it by a logically equivalent conjunction. In this form the inference can now be evaluated by means of truth tables.

Infinite Domains. When we attempt to extend our decision procedure beyond finite domains, however, we find it breaking down. Recall from Section 20 that it was at least theoretically possible to extend the sentence logic decision procedure to infinite domains. If the inference being evaluated was valid or a sentence tautologous for 2^m individuals, with m the number of constituent monadic predicates, then we could be assured that it would be valid or tautologous no matter how large the domain. No such result can be attained for relational predicates, and hence no evaluation is even theoretically possible.

Moreover, when we attempt to extend our decision procedure beyond finite domains by means of truth tables with ∃-constituents a similar result awaits us. Consider, for example, the inference,

6) <u>Someone is the teacher of somebody.</u>
∴ Someone is the student of somebody.

of the form

7) $\dfrac{\exists x\, \exists y\, Rxy}{\exists x\, \exists y\, Sxy}$

This is a valid inference, and yet would be shown invalid by the decision procedure of Section 23.

To see this we first abbreviate the premiss and conclusion by dropping variables to obtain ∃∃R and ∃∃S, and then expand them into their distributive normal forms. The expansion of the premiss and conclusion is

∃∃R   and   ∃∃S
∃∃(RS R$\bar{S}$)   ∃∃(RS ∨ $\bar{R}$S)
∃∃(RS) ∨ ∃∃(R$\bar{S}$)   ∃∃(RS) ∨ ∃∃($\bar{R}$S)

In the expanded premiss and conclusion there are the three ∃-constituents ∃∃(RS), ∃∃(R$\bar{S}$), and ∃∃($\bar{R}$S). The truth table enumerating the 8 possible interpretations of these constituents has the following form:

∃∃(RS)	∃∃(R$\bar{S}$)	∃∃($\bar{R}$S)	∃∃(RS) ∨ ∃∃(R$\bar{S}$)	∃∃(RS) ∨ ∃∃($\bar{R}$S)
T	T	T	T	T
T	T	F	T	T
T	F	T	T	T
T	F	F	T	T
F	T	T	T	T
F	T	F	T	F
F	F	T	F	T
F	F	F	F	F

This valid inference is shown to be invalid under the sixth interpretation where it is assumed that it is false both that someone is both a teacher and a student and that someone is a student and not a teacher, but true that someone is a teacher and not a student. We can exclude this sixth interpretation, to be sure, as a possibility. Since '...is a teacher of...' and '...is a student of...' are the converse of one another, it is impossible for there to be a teacher of someone and there not be a student. But this exclusion cannot be carried out on purely formal grounds. We can exclude the possibility only because of the particular kind of predicates that the predicate schemata R and S of 7) are understood to represent.

We can also attempt again to regard 6) as an enthymeme and add to 7) the premiss $\forall x \forall y (Rxy \equiv Syx)$ specifying R and S are the converse of one another. The resulting inference schema will have a distributive normal form (a much more complex one which must represent the order of variables x and y) with $\exists$-constituents. But again the resulting truth table will have among its possible interpretations those that must be excluded by virtue of the special character of dyadic predicates represented in order to show the inference to be valid.

It should not be concluded that no inference in which dyadic predicates occur can be shown to be valid by the decision procedure of Section 23. For many inferences containing such predicates the type of predicate has no bearing on their evaluation. An example of such inferences would be one of the form

$$\frac{\forall x \forall y (Rxy \vee Sxy)}{\forall x \forall y \neg Rxy}$$
$$\overline{\forall x \forall y Sxy}$$

This can be shown to be valid, since no matter what kind of relation is expressed by the predicates represented by R and S, the premisses will entail the conclusion.

What holds of inferences whose recurring constituents are dyadic predicates holds also for those with triadic, tetradic, and generally n-adic predicates. There will be valid inferences containing these more complex predicates which cannot be shown valid by virtue of their form alone and the logically possible interpretations of their constituents. It might be asked whether there is any other type of procedure that might possibly be used to evaluate inferences and general sentences containing relational predicates. The answer turns out to be in the negative, and is due to a proof by Alonzo Church that no decision procedure of any kind is possible.[1] The fact that for relational predicates the kind of relation expressed may determine the validity of an inference of the logical status of a general sentence in which they occur marks them off from monadic predicates as regards the availability of a decision procedure.

Exercises.

I. Classify the relations expressed by the following sentences with respect to symmetry, transitivity, and reflexivity.
1. Tom is older than Bill.
2. 3 is the square root of 9.
3. 7 is less than 9.
4. The victory occurred at the same place as the defeat.
5. The plague caused the general's death.
6. Boston is to the north of New York.
7. The explosion was simultaneous with the accident.

    8. Mary is the mother of Bill.
    9. Tom is the brother of Bill.
    10. 8 is divisible without remainder by 4.

II. Evaluate by means of truth tables the following inferences as valid or invalid by adding a premiss specifying the type of relation that is expressed.

    1. Tom is the spouse of Mary. Therefore, Mary is the spouse of Tom.

    2. Boston is to the north of New York. New York is to the north of Washington. Hence, Boston is to the north of Washington.

    3. 5 is less than 7. Hence, 7 is greater than 5.

    4. We should conclude that Bill is older than Tom. For suppose that Tom were older than Bill. Then he would be older than Harry. But Tom is not older than Harry.

    5. John is married to either Alice or Mary. But he is not married to Mary unless he is rich. John is not rich. Therefore, Alice is married to John.

25. The Principle of Bivalence

The truth table procedures developed in Chapter II and the present chapter rest on what is called the __principle of bivalence__, the principle that a proposition is __either true or false__. We pause now to examine this assumption and to trace out the consequences that would result from denying it.

The principle of bivalence has been assumed in determining the number of possible interpretations listed in a truth table. By assuming that a proposition can only be either true or false we are assured that for n constituents there are at most 2^n interpretations that are exhaustive of all the possibilities. Many propositions depend for their tautological status on this restriction on the number of possible interpretations. For example, $A \vee -A$, the "law of excluded middle," is a tautology because of the principle. The fact that the proposition expressed by the sentence A can only be either true or false determines along with the definitions of negation and disjunction that $A \vee -A$ is necessarily true. The principle is also assumed in listing the $2^{2^n}-1$ possible interpretations of the $\exists$-constituents generated by n monadic predicates, and is thus a condition for many propositions being tautologies in predicate logic.

__Arguments for denying the principle__. The principle of bivalence seems obvious enough, and yet there have been persuasive arguments given for denying it. Among the most important of these are those based on the undecidability of propositions about the future and on the commitment to determinism of two-valued logic.

1. Undecidability. The argument from the undecidability of future propositions originated with the Polish logician Lukasiewicz.[1] The argument runs roughly as follows. Understanding by a proposition being true or false that it is judged or decided to be true or false (or asserted or denied), it seems that excluded middle does not hold of propositions about the future such as that expressed by 'It will rain tomorrow'. Granted that our present evidence about tomorrow's weather is inconclusive, we shall not be able to decide in the present the truth or falsity of this proposition, nor have grounds for either asserting or denying it. Hence, it is neither true or false. It would seem, then, that to our values of 'true' and 'false' we should add a third value 'undecided' which many propositions about the future should take. This reasoning can be extended to tentative scientific hypotheses, which also seem to require this third value. It can also be extended, it would seem, to certain propositions of mathematics for which we have not been able to determine whether they or their negations are provable from the axioms of a certain theory, e.g. "Fermat's Last Theorem" that there exists numbers x,y,z,n, with $n>2$ and $x,y,z \neq 0$ such that $x^n+y^n=z^n$. It is thus argued that we must reject the principle of excluded middle and replace it by the principle that a proposition must be true, false, or undecided.

2. Indeterminacy. The second argument for denying excluded middle was discussed and apparently advanced by Aristotle.[2] It may be formulated as follows. It seems that to say propositions about future actions are either true or false has the unacceptable consequence of committing us to the philosophic position of determinism. For if a proposition such as that expressed by 'Jones will drive his car tomorrow' is now true, then it would seem that the action it describes must be performed, that Jones must drive his car; while if the proposition is false, the action seems determined not to occur. Now we think that actions such as driving one's car are not determined, that we can freely decide whether or not to perform them. But if this is so, then we must reject the principle of excluded middle and admit that some propositions are neither true nor false, but instead have the value of 'indeterminate'. Only in this way can we avoid the commitment to determinism.

Whether or not to accept these arguments is still a subject of controversy. A major barrier towards their acceptance has been the difficulty of conceiving the kind of values that are proposed as supplementing truth and falsity. As we have seen, one argument against the bivalence principle proposes the undecided as an alternative value, another the indeterminate. But neither seem to provide a genuine contrast with the true and false. For what contrasts with an undecided proposition such as that it will rain tomorrow is a decided proposition, not a true or a false one. Within decided propositions we can distinguish those that have been decided as true and those decided as false. We thus have a trichotomy consisting of propositions decided as true, those decided as false, and those undecided. But this is an epistemic contrast between types of decisions reached about a proposition by human knowers, not a contrast between possible truth values. Similar reasoning applies to an indeterminate proposition, e.g. that Jones may or may not drive his car tomorrow. This contrasts to a determinate proposition, either one stating the necessity of Jones' driving or the impossibility of this event. This contrast is not between truth values of proposition, but instead one between propositions stating the necessity, contingency, or impossibility of a certain event.

By "logical conventionalism" can be understood the position that logical laws are true by convention, depending as they do on principles such as bivalence for which there are alternatives. The choice among these alternatives is dictated, according to the conventionalists, by the type of problem to which our logic is being applied. For certain purposes 2-valued logic may be preferred, but for others, e.g. in applying logic to quantum mechanics or proof construction in mathematics, some form of 3-valued logic may by more useful. The issue involved in assessing the arguments against the bivalence principle is thus the truth or falsity of the conventionalist thesis.[3]

A Three-Valued Logic System. Let us not attempt to resolve this controversy here, but instead trace out the consequences of admitting a third value called the 'undecided' or the 'indeterminate' in addition to the 'true' and 'false'. It has been shown possible to construct decision procedures analogous to those we have been using. This is done by first giving truth table definitions of the logical connectives relative to three values and then applying these definitions in a decision procedure. There are a variety of three-valued systems. We shall present one first proposed by Lukasiewicz.[4]

Let us represent the third value, the undecided or the indeterminate, by 'I'. Then the truth values for the negation of A relative to the values for A may be given by the following table:

A	-A
T	F
I	I
F	T

The value I for -A is given, since if A is indeterminate or undecided so would be its negation. The truth tables defining the other connectives may be more easily presented by adopting a matrix arrangement in which the intersection of rows and columns indicates the truth value of a molecular schema relative to the values of its constituents. The truth tables for conjunction and disjunction are

	∧	T	I	F
	T	T	I	F
A	I	I	I	F
	F	F	F	F

and

	∨	T	I	F
	T	T	T	T
A	I	T	I	I
	F	T	I	F

The connective being defined is indicated in the upper-left corner of the truth tables. The possible values of A and B are given in the left column and top row respectively. The definitions given seem to accord with our intuitions for conjunction and (inclusive) disjunction when extended to three values. A conjunction is false when one conjunct is false, but indeterminate (undecided) when a conjunct is indeterminate. A disjunction is indeterminate except where a disjunct is true or both disjuncts false.

Less intuitively obvious are the truth tables given by Lukasiewicz for material implication and equivalence.

	⊃	T	I	F
	T	T	I	F
A	I	T	T	I
	F	T	T	T

	≡	T	I	F
	T	T	I	F
A	I	I	T	I
	F	F	I	T

Under the first definition A⊃B is undecided (indeterminate) if A is true and B undecided. When A and B are both undecided, on the other hand, A⊃B is true. But this seems contrary to what we should expect, since if A and B are both undecided, then A may turn out true and B false. The implication thus seems to be possibly false and deserve the third value of I. The definition also has the effect of not allowing us to define implication in terms of negation and conjunction or negation and disjunction in the usual way, since the truth tables of -(A∧-B) and -A∨B are different from that given above for implication.

The definition given of implication, however, does have the merit of allowing us a decision procedure for evaluating whether schemata express tautologies. We simply construct a truth table with the possible interpretations of the constituent schemata for three values, and if a certain schema is true for all of them it expresses a tautology. Thus, the following truth tables show that A⊃A is a tautology in 3-valued logic, but that the law of excluded middle A∨-A is not.

A	A⊃A		A	A∨-A
T	T		T	T F
I	T		I	I I
F	T		F	T T

It thus can be seen how the law of excluded middle presupposes the principle of excluded middle; deny the principle and the law no longer is a logical truth. Similarly, it can be shown that the law of contraposition A⊃B≡-B⊃-A is a tautology, while the 2-valued tautology [(A⊃B)⊃A]⊃A fails to be tautologous. The truth table for the latter is the following:

A	B	[(A⊃B) ⊃ A] ⊃ A
T	T	T T T
T	I	I T T
T	F	F T T
I	T	T T I
I	I	T I T
I	F	I T I
F	T	T F T
F	I	T F T
F	F	T F T

In general, for n atomic schemata there will be 3^n possible interpretations if we assume three values. If a given molecular schema is true for all of them it expresses a tautology; otherwise it is either contingent or contradictory.

This procedure can, of course, be extended to predicates by regarding a given ∃-constituent ∃xPx as either true, false, or indeterminate. Here again there will be tautologies in 2-valued logic that are contingent when we add a third value. An example of one is ∀x-Px ∨ ∃xPx whose normal form is -∃xPx ∨ ∃xPx. Like the law of excluded middle the proposition expressed by a sentence of this form will have the value indeterminate when its single ∃-constituent ∃xPx has the value of indeterminate. If we are to again exclude the possibility of the empty domain in interpreting ∃-constituents, we must exclude interpretations in which an ∃-constituent is indeterminate while the others are all false in addition to those in which all ∃-constituents are false. With this exclusion the schema ∀xPx ⊃ ∃xPx remains a tautology.

NOTES FOR CHAPTER V

Section 20.

1. Methods for abbreviating this truth table procedure have been devised. For an example of one see Jeffrey [Logic], Ch. 6, where his truth tree method is extended by adding rules for the quantifiers. Even with these procedures evaluation is usually very lengthy and complex.

2. For a version of this proof originally due to Bernays and Schonfinkel see Ackermann [Decision Problem], Chs. V and VI.

Section 21.

1. For this argument see Strawson [Logical Theory], pp. 163-179. As Strawson points out, besides sub-alternation the three basic relations of the Square of Opposition also no longer hold for the A, E, I, O forms when represented by the standard symbolism of predicate logic. By representing the A form by $\forall x(Sx \supset Px) \land \exists xSx$ and the E form by $\forall x(Sx \supset -Px) \land \exists xSx$, however, sub-alternation and the contrary, sub-contrary, and contradictory relations do hold.

2. Hence the discrepancy between the number of syllogisms accepted as valid in the traditional Aristotelian logic and the number that can be evaluated as valid by Venn diagrams. 15 of the 256 possible syllogisms are valid by Venn diagrams. They are AAA-I, EAE-I, AII-I, EIO-I, EAE-II, AEE-II, EIO-II, AOO-II, IAI-III, AII-III, OAO-III, EIO-III, AEE-IV, IAI-IV, EIO-IV. In addition to these nine additional syllogisms were accepted as valid in the traditional logic: AAI-I, EAO-I, AEO-II, EAO-II, AII-III, EAO-III, AEO-IV, EAO-IV, AAI-IV. The validity of these nine syllogisms requires the assumption of the existential commitment of the universal forms. Cf. Quine [Methods], Sec. 14.
For the most complete study of Aristotelian logic see Lukasiewicz [Syllogistic].

Section 22.

1. The version that follows of a procedure originally due to Herbrand is based on that in von Wright [Logical Studies], Essays I and II. Another version using his method of reduction can be found in Quine [Methods], Secs. 17-21.

Section 23.

1. The term 'tautology' was originally used by Wittgenstein in [Tractatus] with reference only to propositions necessarily

true for all possible interpretations of their atomic constituents, i.e. the logical truths of sentence logic and those of predicate logic when expressed as truth functions of atomic propositions. The extension of the term to logical truths of predicate logic as truth functions of $\exists$-constituents is due to von Wright [Logical Studies], Essays I and II. Such propositions true in every non-empty domain are usually referred to as "universally valid" propositions in the literature.

Section 24.

1. The proof is in Church [Entscheidungsproblem].

Section 25.

1. See Lukasiewicz [Logic], pp. 77, 78.

2. See the Kneales [Development], pp. 48ff. and Hintikka [Sea Fight] for discussions of Aristotle's version of the argument.

3. The conventionalist position is stated by Carnap [Syntax], Sec. 17. For a criticism of it see the Kneales [Development], pp. 629-651.

4. For this system see Lukasiewicz [Logic], pp. 77-83. For a survey of the various three-valued systems see Rescher [Many-Valued Logic]. Three-valued systems have been generalized by Lukasiewicz and others to n-valued systems, where n can take any value between 1 (true) and 0 (false).

VI. THE PREDICATE CALCULUS

26. Rules for the Universal Quantifier

In Chapter III we developed a calculus in which interconnections between rules governing inferences of the form evaluated in sentence logic were traced. This calculus also provided an alternative means of establishing the validity of such inferences. In the present chapter the sentence calculus is extended in order to trace the interconnections between rules governing inferences of the form evaluated in predicate logic. The resulting extension is the predicate calculus. As before, the predicate calculus provides us with an alternative, and very much simpler, means of establishing the validity of inferences and the fact that a given sentence expresses a tautology. Inferences and general sentences that can only be shown to be valid and to express tautologies after a lengthy and tedious enumeration of possibilities by means of the decision procedures outlined in the previous chapter can easily be verified by the predicate calculus. Perhaps even more importantly, inferences containing relational predicates for which no decision procedure is available can be shown to be valid by means of the calculus.

The predicate calculus is developed by adding to the Principle of Reiteration and primitive rules R1-R10 of the sentence calculus rules governing the universal and existential quantifiers. In addition, the calculus requires two more syntactic principles in order to apply the quantifier rules that are added. The first of these principles is the

Principle of Restricted Reiteration: No schema in which a free variable occurs may be reiterated into a sub proof general with respect to that variable.

By a general sub proof we mean a sub proof whose conclusion is proved independently of any particular individual in the domain over which its variable ranges. We indicate that a sub proof is general with respect a certain free variable by placing the variable to the left of it, as in

```
1 | ∃xPx          p
2 | Px            p
3 |   | ∃xPx      1,r
4 | x | Px        2,r   (forbidden)
  |   | .
  |   | .
  |   | .
```

The Principle of Restricted Reiteration forbids reiteration of expressions in which this variable occurs free into the sub proof. Thus, step 3 is permitted by the Principle, but not step 4 in which the open schema Px is reiterated. Restricting reiteration in this manner avoids invalid inferences that would otherwise result. Notice that there is no premiss in the general sub proof of the above example. This occurs when such proofs are used in applying the first rule governing the universal quantifier.

This first rule is

R11. <u>Universal Quantifier Introduction</u>: <u>From a proof within a general sub proof of an open schema we may infer its universal generalization</u>.

This rule permits an inference such as

$$\begin{array}{|l} x \begin{array}{|l} \vdots \\ Px \supset Qx \end{array} \\ \forall x(Px \supset Qx) \end{array}$$

where we assume we have proved $Px \supset Qx$ within the general sub proof. Given this proof, R11 permits the inference of its universal generalization $\forall x(Px \supset Qx)$. The rule can be interpreted as permitting the inference from a predicate holding of any individual in a given domain to its holding of every individual. If P implies Q holds for any arbitrary individual we happen to choose, then it clearly holds for every individual. The symbolic formulation of the rule of universal quantifier introduction requires our introducing the expression '$\phi\nu$' as a meta-schema representing any open schema, with the symbol 'ν' representing the variable occurring free in the schema. For example, $\phi\nu$ may stand for '$Px \supset Qx$' with 'x' the free variable, or for 'Pz' with free 'z'. To represent the universal generalization of an open schema we use the expression '$\forall\mu\phi\mu$' in which 'μ' represents the bound variable replacing the free variable of the open schema at every occurrence. The symbolic formulation of R11 is thus

$$\begin{array}{c} \nu \begin{array}{|l} \vdots \\ \phi\nu \end{array} \\ \hline \forall\mu\phi\mu \end{array}$$

with $\phi\nu$ representing the open schema proved in the sub proof general with respect to some variable ν and $\forall\mu\phi\mu$ its universal generalization.

These meta-schemata are also used in symbolizing the second rule for the universal quantifier.

R12. <u>Universal Quantifier Elimination</u>: From <u>a universal quantificational schema</u> $\forall\mu\phi\mu$ <u>we may infer an open schema</u> $\phi\nu$ <u>in which the same free variable</u> ν <u>replaces the bound variable</u> μ <u>at every occurrence</u>.

The symbolic formulation of R12 is

$$\frac{\forall\mu\phi\mu}{\phi\nu}$$

with $\phi\nu$ representing the open schema inferred from the universal quantificational schema in which the free variable represented by ν replaces the bound variable μ at every occurrence. With this rule we could license the inference from $\forall x(Px \supset Qx)$ to $Py \supset Qy$, where we drop the quantifier and replace the bound occurrences of x by free occurrences of y.

The quantifier rules just formulated are used in conjunction with the rules of the rules of the sentence calculus in conducting proofs. Both quantifier rules are applied in the following conditional proof.

```
1)   1 | ∀x(Px ⊃ Qx)         p
     2 | ┌ ∀x(Px ⊃ Qx)       1,r
     3 | x| Px ⊃ Qx           2, ∀ elim
     4 | └ -Qx ⊃ -Px          3, contrapos
     5 | ∀x(-Qx ⊃ -Px)       2-4, ∀ intro
```

The proof of the conclusion proceeds by first proving the open schema $-Qx \supset -Px$ within a general sub proof. Given this proof the rule of universal quantifier introduction allows us to infer the conclusion. To prove $-Qx \supset -Px$ we reiterate step 1 into the general sub proof. Since the free variable x does not occur in the premiss, reiteration is permitted by the Principle of Restricted Reiteration. Step 3 is inferred from 2 by the rule of universal quantifier elimination, with the same variable x being used as the free variable in the open schema that is the bound variable in the quantificational schema. The conclusion of the sub proof follows by the derived rule of contraposition of the sentence calculus, with the meta-schemata ϕ and ψ of the rule now representing open schemata. This proof illustrates the basic function of the quantifier rules, that of first permitting the quantifiers to be removed in order for the open schema to be transformed according to the rules of the sentence calculus and then permitting the quantifiers to be reintroduced.

The reiteration of a quantificational schema into a general sub proof and then the dropping of the quantifier occurs so frequently within proofs that it is convenient to combine the two steps. The following proof illustrates this.

2)
```
1 | ∀x(Qx ⊃ Rx)          p
2 |    | Px ∧ Qx          p
3 |    | Qx               2, ∧ elim
4 | x  | Qx ⊃ Rx          1,r, ∀ elim
5 |    | Rx               3,4, m.p.
6 |    Px ∧ Qx ⊃ Rx       2-5, ⊃ intro
7 | ∀x(Px ∧ Qx ⊃ Rx)     2-6, ∀ intro
```

In step 4 in the proof we both reiterate the premiss of the main proof and eliminate the quantifier, combining two steps into one. We shall follow this practice in future proofs. Notice that we apply the Implication Strategy of the sentence calculus in order to prove the schema Px ∧ Qx ⊃ Rx in step 6. Once having proved this open schema, the conclusion can be inferred by the rule of ∀ intro.

The strategy used in conducting proofs 1) and 2) is one that has general application.

Universal Quantifier Strategy: If the conclusion is of the form of a universal quantificational schema, then set up a general sub proof whose conclusion is the open schema within the scope of the universal quantifier. If this schema can be proved, the conclusion can be inferred by the rule of universal quantifier introduction.

In proof 1) we would use this strategy by constructing a general sub proof with -Qx ⊃ -Px as its conclusion. Proving this open schema allows us to infer the conclusion by ∀ intro. The same strategy is used in proof 2) in constructing the general sub proof of steps 2-6.

The rule of universal quantifier elimination can also be adapted for use in proofs in which occur individual schemata. An example of such a proof is

3)
```
1 | ∀x(Px ⊃ Qx)     p
2 | Pa              p
3 | Pa ⊃ Qa         1, ∀ elim
4 | Qa              2,3, m.p.
```

in which we infer Pa ⊃ Qa from ∀x(Px ⊃ Qx) by ∀ elim in order to derive the conclusion Qa by <u>modus ponens</u>. This requires us to understand the rule of ∀ elim as permitting us to substitute any arbitrary individual schema for the free variables of the open schema inferred from the quantificational schema. When interpreted the rule permits us to infer from a predicate holding of every individual in a domain that it holds also of any particular individual. This is clearly justified in terms of the equivalence between universal sentences and conjunctions of atomic sentences noted in Section 20. If ∀xPx is equivalent to $Pa_1 \wedge Pa_2 \wedge \ldots \wedge Pa_n$, then because from a conjunction follows any conjunct we can infer any of the atomic sentences Pa_1, Pa_2, ..., Pa_n. Adapted to apply to individual schemata the rule of $\forall$ elim becomes the series of rules,

$$\frac{\forall \mu \phi \mu}{\phi a} \quad , \quad \frac{\forall \mu \phi \mu}{\phi b} \quad , \quad \frac{\forall \mu \phi \mu}{\phi c} \quad , \quad \ldots$$

where individual schemata a,b,c,... are substituted for the free variables of the open schema $\phi \nu$ inferred from $\forall \mu \phi \mu$. The same individual schema must be substituted for every occurrence of the free variable.

There are, in fact, two entirely different applications of the rule of universal quantifier elimination. Within a general sub proof the rule is usually used to infer an open schema with free variables from the quantificational schema. Within any proof not a general sub proof, on the other hand, the rule is invariably used to infer a schema in which some arbitrary individual schema is substituted at every occurrence of the free variable in the open schema. Every schema outside a general sub proof must represent a sentence of the object language capable of being interpreted as expressing a true or false proposition. Hence the requirement for substituting for free variables to form schemata representing closed sentences. In contrast, general sub proofs enable us to temporarily suspend the interpretation of expressions prior to reintroducing quantifiers. Here we allow the occurrence of schemata representing expressions incapable of being interpreted as true or false. The two different applications of the rule are regulated by the following

<u>Principle of Substitution</u>: <u>An individual schema must be substituted for a free variable within any proof not a general sub proof for every occurrence of that variable, but substitution for free variables within general sub proofs is not required</u>.

This syntactic principle allows us to infer in proofs 1) and 2) from a quantificational schema to an open schema, since the inferences take place within general sub proofs. In proof 3), on the other hand, the elimination of the quantifier takes place within the main proof, and we thus substitute an individual schema for occurrences of the free variable.

<u>Exercises</u>. Prove the following.

1. $\forall x(Px \supset Qx) \vdash \forall xPx \supset \forall xQx$ 7. $\forall x(Px \lor Qx), \forall x\text{-}Px \vdash \forall xQx$

2. $\forall xPx \lor \forall xQx \vdash \forall x(Px \lor Qx)$ 8. $\forall x(Px \land Qx \supset Rx) \vdash \forall x[Px \supset (Qx \supset Rx)]$

3. $\forall xPx \land \forall xQx \vdash \forall x(Px \land Qx)$ 9. $\forall x[Px \supset (Qx \supset Rx)] \vdash \forall x(Px \land Qx \supset Rx)$

4. $\forall x(Px \supset Qx) \vdash \forall x(Px \supset Qx \lor Rx)$ 10. $\forall x(Px \supset Qx), -Qb \vdash -Pb$

5. $\forall x(Px \equiv Qx) \vdash \forall xPx \equiv \forall xQx$ 11. $\forall x(Px \land Qx), -Pa \vdash Rb$

6. $\forall xPx, \forall x(Px \supset Qx) \vdash \forall xQx$ 12. $\forall x(Px \supset Qx) \vdash Pc \supset (Rc \supset Qc)$

27. Rules for the Existential Quantifier

Introduction and elimination rules can also be formulated for the existential quantifier.

R13. **Existential Quantifier Introduction:** <u>From an open schema $\phi\nu$, we may infer its existential generalization of the form $\exists\mu\phi\mu$, where the bound variable μ replaces ν (and only ν) at every occurrence.</u>

$$\frac{\phi\nu}{\exists\mu\phi\mu}$$

The rule thus states that if a predicate holds of any arbitrary individual then it holds of at least one. Applying the Principle of Substitution to this rule, it is understood that within a proof not a general sub proof some arbitrary individual schema will be substituted for every occurrence of the free variable ν within the open schema $\phi\nu$. In this case the rule now states that if a particular individual is ϕ, then at least one thing is ϕ. This can be justified by the equivalence between an existential sentence and a disjunction of atomic sentences, since a disjunction is entailed by any one of its disjuncts. Like the rule of $\forall$ elim this second form of $\exists$ intro becomes equivalent to the indefinite number of rules,

$$\frac{\phi a}{\exists\mu\phi\mu} \quad , \quad \frac{\phi b}{\exists\mu\phi\mu} \quad , \quad \frac{\phi c}{\exists\mu\phi\mu} \quad , \quad \ldots$$

R14. **Existential Quantifier Elimination:** <u>From an existential schema $\exists\mu\phi\mu$ and a general sub proof showing that the open schema $\phi\nu$ (in which only occurrences of μ are replaced by ν) leads to a schema ψ in which the free variable ν does not occur we may infer ψ.</u>

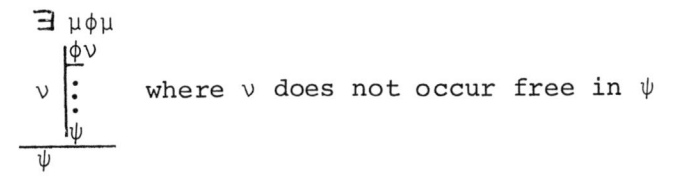

$$\begin{array}{l} \exists\mu\phi\mu \\ \quad \begin{vmatrix} \phi\nu \\ \vdots \\ \psi \end{vmatrix} \\ \hline \psi \end{array} \quad \text{where } \nu \text{ does not occur free in } \psi$$

The restriction that the free variable ν not occur in ψ prevents the following proof being possible.

```
1)    1 | ∃xPx         p
      2 |   | Px       p
      3 | x | Px       2,r
      4 | Pa           1,2-3, ∃ elim (forbidden)
```

Since a free variable is not allowed in a main proof, we substitute an individual schema for x in step 4. But the resulting inference is clearly invalid: because something is P it does not follow that some one individual is (someone is President of the U. S., but it does not follow that George Wallace is). We avoid such a fallacious inference by excluding the possibility of step 4. Since the free variable x occurs in step 3, it cannot be brought into the main proof and substituted for.

The rule of ∃ elim is admittedly an extremely artificial rule and difficult to intuitively justify.[1] The best that can be easily said for it is that it proves to be a rule which allows us to eliminate the existential quantifier from a quantificational schema, transform the open schema within the scope of the quantifier according to the rules of the sentence calculus, and then reintroduce existential quantifiers in such a way as to never infer from true premisses a false conclusion. The restrictions on reiteration are such as to insure that only existential schemata can be proved as the conclusion of the general sub proof to which the rule is applied. Hence only existential conclusions can be inferred by the rule from existential premisses.

The following proof illustrates an application of the two existential quantifier rules.

```
2)    1 | ∃x(Px ∨ Qx)             p
      2 |   | Px ∨ Qx             p
      3 |   |   | Px              p
      4 |   |   | ∃xPx            3, ∃ intro
      5 |   |   | ∃xPx ∨ ∃xQx     4, ∨ intro
      6 | x |   | Qx              p
      7 |   |   | ∃xQx            6, ∃ intro
      8 |   |   | ∃xPx ∨ ∃xQx     7, ∨ intro
      9 |   | ∃xPx ∨ ∃xQx         2,3-5,6-8, ∨ elim
     10 | ∃xPx ∨ ∃xQx             1,2-9, ∃ elim
```

Here we infer our conclusion from the existential schema as premiss by showing that the open schema within the scope of the quantifier, Px ∨ Qx, leads to the conclusion within a general sub proof of steps 2-9. ∃xPx ∨ ∃xQx is proved to follow from Px ∨ Qx by means of the rule of ∨ elim of the sentence calculus, with use of the rule of ∃ intro in steps 4 and 7. Since ∃xPx ∨ ∃xQx does not contain the variable x free, it may be brought into the main proof in accordance with ∃ elim. In general, whenever there is given an existential schema from which we want to derive a conclusion we will need to eliminate the quantifier by the rule of ∃ elim. We thus have another strategy.

Existential Quantifier Strategy: If it is desired to prove a conclusion from a given existential quantificational schema, then use the rule of ∃ elim by setting up a general sub proof whose premiss is the open schema within the scope of the existential quantifier and whose conclusion is the conclusion to be proved.

We would construct proof 2) with this strategy by first setting up the following outline of a proof.

```
1   ∃x(Px ∨ Qx)           p
2      Px ∨ Qx             p
  x
  .     .
  .     .
n-1    ∃xPx ∨ ∃xQx
n   ∃xPx ∨ ∃xQx           1,2-n-1, ∃ elim
```

and then proceeding to derive step n-1 from 2 in the manner shown in 2).

Rules for both the existential and universal quantifiers can be combined within a proof, as illustrated in the following example:

```
3)  1   ∀x(Px ⊃ Qx)        p
    2   ∃x(Rx ∧ Px)        p
    3      Rx ∧ Px          p
    4      Px               3, ∧ elim
    5      Px ⊃ Qx          1,r, ∀ elim
    6   x  Qx               4,5, m.p.
    7      Rx               3, ∧ elim
    8      Rx ∧ Qx          7,6, ∧ intro
    9      ∃x(Rx ∧ Qx)      8, ∃ intro
   10  ∃x(Rx ∧ Qx)         2,3-9, ∃ elim
```

Since the existential schema ∃x(Rx ∧ Px) is a premiss, we use the Existential Quantifier Strategy and construct the general sub proof leading from the open schema Rx ∧ Px to the conclusion.

The necessity of the Principle of Restricted Reiteration becomes apparent for proofs involving both quantifiers. Without it the following proof would be possible:

```
4)  1 | ∃xPx            p
    2 |   | Px          p
    3 | x | x | Px      2,r (forbidden)
    4 |   | ∀xPx        3, ∀ intro
    5 | ∀xPx            1,2-4, ∃ elim
```

Because Px is reiterated into a sub proof general with respect to x, step 3 is forbidden by the Principle of Reiteration.

As before for the sentence calculus (cf. Section 14), every conditional proof within the predicate calculus is the proof of a derived rule of inference that can be applied to future proofs. Let us call the negation of an open schema within the scope of a quantifier, e.g. ∀x-Px, the _internal negation_ form of a quantificational schema, while the negation of a quantificational schema, e.g. -∀xPx, we shall call its _external negation_. Two especially useful derived rules are those allowing us to infer from the internal negation form ∀x-Px to the external negation existential form -∃xPx and from the latter to the former. The proofs of these are as follows:

```
1 | ∀x-Px          p              1 | -∃xPx            p
2 |  | ∃xPx        p              2 |   | Px           p
3 |  |  | Px       p              3 |   | | ∃xPx       2, ∃ intro
4 |  | x| Px       3,r            4 | x | | -∃xPx      1,r
5 |  |  | -Px      1,r ∀ elim     5 |   | -Px          2-4, -intro
6 |  |  | -∀x-Px   4,5, -elim     6 | ∀x-Px            2-5, ∀ intro
7 |  | -∀x-Px      2,3-6, ∃ elim
8 | ∀x-Px          1,r
9 | -∃xPx          2-8, -intro
```

The rules ∀μ-φμ ⊢ -∃μφμ and -∃μφμ ⊢ ∀μ-φμ just proven we can refer to as the rules of "negation existential quantifier introduction" (-∃ intro) and "negation existential quantifier elimination" (-∃ elim). They are listed in Appendix I as derived rules DR9 and DR10.

Recall from the end of Section 23 that there are inferences whose validity depends on the recurrence of both predicates and sentences as wholes. Proofs in which occur both sentence and predicate schemata can be constructed with our presently available rules. An example of one follows.

```
 1 | ∀x(A ∨ Px)         p
 2 |  | -(A ∨ ∀xPx)     p
 3 |  | -A ∧ -∀xPx      2, -∨ elim
 4 |  | -A              3, ∧ elim
 5 |  |  | A ∨ Px       1,r, ∀ elim
 6 |  | x| -A           4,r
 7 |  |  | Px           5,6, d.s.
 8 |  | ∀xPx            5-7, ∀ intro
 9 |  | -∀xPx           3, ∧ elim
10 |  | --(A ∨ ∀xPx)    2-9, -intro
11 | A ∨ ∀xPx           10, --elim
```

The proof is constructed in accordance with Negation Strategy b), with $\forall x Px$ and $-\forall x Px$ the contradiction proved within the sub proof. Notice how in step 5 the rule of $\forall$ elim is applied to the schema $\forall x(A \vee Px)$. The scope of the quantifier is here regarded as the schema $A \vee Px$ in which x occurs bound. By applying the rule of $\forall$ elim we drop the quantifier, leaving x free. The derived rule just proven, $\forall \mu(\psi \vee \phi\mu) \vdash \psi \vee \forall \mu \phi \mu$, is listed in the Appendix along with $\exists \mu(\psi \vee \phi\mu) \vdash \psi \vee \exists \mu \phi \mu$ which is to be proved in the exercises that follow.

Exercises. Prove the following. Introduce and use derived rules based on previous proofs where convenient.

1. $\exists x(Px \wedge Qx) \vdash \exists x(Px \vee Qx)$
2. $\exists x(Px \wedge Qx) \vdash \exists x Px \wedge \exists x Qx$
3. $\exists x Px \vee \exists x Qx \vdash \exists x(Px \vee Qx)$
4. $Pa \supset Qa, -Qa \vdash \exists x -Px$
5. $\exists x(Px \supset Qx) \vdash \forall x Px \supset \exists x Qx$
6. $\forall x(Px \supset Qx) \vdash \exists x Px \supset \exists x Qx$
7. $\exists x Px \supset \forall x Qx \vdash \forall x(Px \supset Qx)$
8. $\forall x(Px \equiv Qx) \vdash \exists x Px \equiv \exists x Qx$
9. $\exists x Px, \forall x Qx \vdash \exists x(Px \wedge Qx)$
10. $\forall x(Px \supset -Qx), \exists x(Rx \wedge Px) \vdash \exists x(Rx \wedge -Qx)$
11. $\forall x(Px \supset Qx), \exists x Px \vdash \exists x(Px \wedge Qx)$
12. $\exists x -Px \vdash -\forall x Px$
13. $-\forall x Px \vdash \exists x -Px$
14. $\forall x Px \supset \exists x Qx \vdash \exists x(Px \supset Qx)$
15. $\exists x Px \supset \exists x Qx \vdash \exists x(Px \supset Qx)$
16. $A \vee \forall x Px \vdash \forall x(A \vee Px)$
17. $\forall x(A \wedge Px) \vdash A \wedge \forall x Px$
18. $A \wedge \exists x Px \vdash \exists x(A \wedge Px)$
19. $\exists x(A \supset Px) \vdash A \supset \exists x Px$
20. $\forall x(Px \supset A) \vdash \exists x Px \supset A$

28. Applications of the Predicate Calculus

Evaluation of Inferences. The predicate calculus is applied to inferences of an object language in order to show them valid in exactly the same manner as for the sentence calculus (cf. Section 14). Every inference in which recur predicates can also be regarded as formulated in accordance with a rule of inference. As an example, we take again the Aristotelian syllogism,

1) All men are mortal.
   All Greeks are men.
   ∴ All Greeks are mortal.

The rule followed in formulating the inference is clearly

> 2) $\forall x(Px \supset Qx)$
> $\underline{\forall x(Rx \supset Px)}$
> $\forall x(Rx \supset Qx)$

The syllogism 1) can be shown to be valid if this rule is valid, one which if it were followed would never allow us to infer a false conclusion from true premisses. Now each of the primitive rules we have adopted for the predicate calculus is a valid rule; the precautions taken to avoid fallacies insures this. If this is so, then by exactly the same reasoning as was used in Section 14 for the sentence calculus a conditional proof that the conclusion of rule 2) follows from its premisses is a proof that the rule is a derived rule of the calculus and hence is also valid. This conditional proof can easily be constructed.

> 3)   1 | $\forall x(Px \supset Qx)$ p
>   2 | $\forall x(Rx \supset Px)$ p
>   3 | $Px \supset Qx$ 1,r, $\forall$ elim
>   4   x | $Rx \supset Px$ 2,r, $\forall$ elim
>   5 | $Rx \supset Qx$ 4,3, h.s.
>   6 | $\forall x(Rx \supset Qx)$ 3-5, $\forall$ intro

The Aristotelian syllogism 1) has therefore been shown to be formulated in accordance with a derived rule of the calculus and to be valid.

    It is clear that the same procedure can be applied to any inference of the form considered in predicate logic. To show that the inference is valid we simply formulate the rule governing it. If we can construct a conditional proof that the rule is a derived rule of the predicate calculus, then we know the inference is valid. This procedure cannot, however, enable us to show that an inference is invalid (cf. Section 14). Failure to construct a proof of a derived rule can either mean that no proof is possible (since the rule is invalid) or that we have been unable to discover a proof that is in principle possible. We have no means within the calculus itself of demonstrating which of the alternatives is applicable. For an effective decision procedure allowing us to determine both validity and invalidity we must utilize the decision procedures of the previous chapter.

    <u>Categorical Proofs</u>. Categorical proofs of a conclusion relative to no premisses in the main proof are constructed exactly as for the sentence calculus. The following is an example of one.

4)
```
1 |  | Px ∧ -Px        p
2 |  | Px             1, ∧ elim
3 | x| -Px            1, ∧ elim
4 |  |-(Px ∧ -Px)     1-3, -intro
5 ∀x-(Px ∧ -Px)      1-4, ∀ intro
```

The proof is constructed by employing the Universal Quantifier Strategy to the conclusion, and then the Negation Strategy a) to the open schema of the general sub proof.

As before, categorical proofs such as 4) provide an alternative (and usually much simpler) means for showing that a given sentence expresses a tautology. If the sentence when represented by a quantificational schema is capable of a categorical proof, then we know that it expresses a tautology because of the validity of our primitive rules. Proof 4) would therefore show that the sentence 'Everything is not both red and not red' expresses a tautology of predicate logic, since the schema representing the sentence is the conclusion of the categorical proof.

Relational Predicates. So far we have restricted application of the predicate calculus to proofs involving monadic predicates. It can also be used to conduct proofs in which two-place or dyadic predicates occur. For example, step 2 of the proof,

5)
```
1 | ∀x ∃y Rxy      p
2 | ∃y Ray         1, ∀ elim
3 | ∃z ∃y Rzy      2, ∃ intro
```

can be licensed by the rule of ∀ elim, with the individual schema a being substituted in accordance with the Principle of Substitution. In step 3 we employ the rule of ∃ intro. Both inferences are enabled by the generality with which the rules of the predicate calculus are stated. The quantifiers in the rules stand for the left-most quantifier of an expression in which may occur more than one quantifier. Thus ∀μφμ is regarded as representing ∀x∃yRxy in the first step of 5), with ∀μ standing for '∀x' and φμ for the expression '∃yRxy' with bound 'x'.

The following conditional proof illustrates the manner in which the calculus is applied to dyadic predicates.

6)
```
1 | ∃x ∀y Rxy              p
2 |   ∃x ∀y Rxy            1, r
3 |     | ∀y Rxy           p
4 | y | x | Rxy            3, ∀ elim
5 |     | ∃x Rxy           4, ∃ intro
6 |   ∃x Rxy               2, 3-5, ∃ elim
7 | ∀y ∃x Rxy              2-6, ∀ intro
```

The proof is constructed by first applying the Universal Quantifier Strategy and then the Existential Quantifier Strategy. The successive stages of the proof are thus

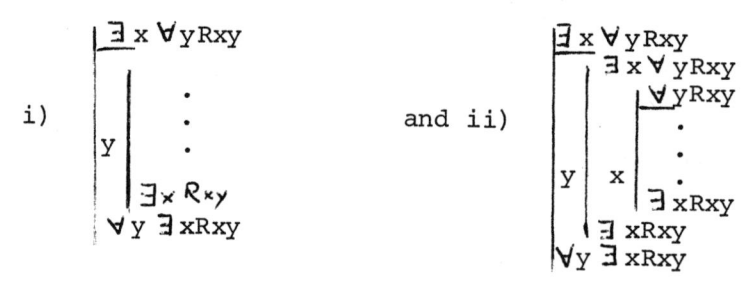

i)
```
| ∃x∀yRxy
|   .
| y .
|   .
| ∃x Rxy
| ∀y ∃xRxy
```

and ii)
```
| ∃x∀yRxy
|  | ∃x∀yRxy
|  |  | ∀yRxy
|  |  |   .
| y| x|   .
|  |  | ∃xRxy
|  | ∃xRxy
| ∀y ∃xRxy
```

It remains only to fill in the steps in ii) leading from $\forall yRxy$ to $\exists xRxy$. Since x does not occur free in $\exists xRxy$ it is permissible to infer it as the conclusion of the first sub proof in accordance with the rule of $\exists$ elim.

In similar fashion the rules of the calculus can be applied to n-place or n-adic predicates generally. Invariably in such proofs successive applications of the elimination rules first derive an open schema without quantifiers which can be transformed by the rules of the sentence calculus. The introduction rules are then applied in such a way as to derive the desired conclusion. The following is an example of a proof involving a triadic predicate and three quantifiers.

```
1  | ∃x∀y∀z(Px ∧ Rxyz)              p
2  |   | ∃x∀y∀z(Px ∧ Rxyz)          1,r
3  |   |   | ∀y∀z(Px ∧ Rxyz)        p
4  |   |   | Px ∧ Rxyz              3, ∀ elim, ∀ elim
5  |   | x | Rxyz                   4, ∧ elim
6  | z y   | Px                     4, ∧ elim
7  |   |   | Rxyz ∧ Px              5,6, ∧ intro
8  |   |   | ∃x(Rxyz ∧ Px)          7, ∃ intro
9  |   | ∃x(Rxyz ∧ Px)              2,3-8, ∃ elim
10 | ∀y ∃x(Rxyz ∧ Px)               2-9, ∀ intro
11 |∀z ∀y ∃x(Rxyz ∧ Px)             2-10, ∀ intro
```

Note that in step 4 we combine steps by two applications of ∀ elim. The proof is constructed by successive applications of the Universal and Existential Quantifier Strategies.

To apply the calculus to show that inferences in which dyadic predicates occur are valid poses special problems. Recall from Section 24 that different types of relations are expressed by dyadic predicates. We can classify these predicates according to the reflexivity, symmetry, and transitivity of the relations they express and determine whether two relations are the converse of one another.

Without specifying the type of dyadic predicate it is often impossible to show that an inference in which it occurs is valid. Consider, for example, the inference,

> 7) Tom is taller than Bill.
> Bill is taller than Jim.
> ∴ Tom is taller than Jim.

It is impossible to show 7) to be valid by showing that the rule governing it can be derived within the calculus. By adding a premiss, however, that specifies that the dyadic predicate '...is taller than...' expresses a transitive relation 7) can be shown to be valid. The inference then becomes

> 8) Tom is taller than Bill.
> Bill is taller than Jim.
> If a first is taller than a second and that second is taller than a third, then the first is taller than the third.
> ∴ Tom is taller than Jim.

The rule governing 8) is now

> 9) Rab
> Rbc
> $\forall x \forall y \forall z (Rxy \land Ryz \supset Rxz)$
> Rac

The proof that it is a derived rule is

> 10)
> 1. Rab — p
> 2. Rbc — p
> 3. $\forall x \forall y \forall z (Rxy \land Ryz \supset Rxz)$ — p
> 4. $\forall y \forall z (Ray \land Ryz \supset Raz)$ — 3, $\forall$ elim
> 5. $\forall z (Rab \land Rbz \supset Raz)$ — 4, $\forall$ elim
> 6. $Rab \land Rbc \supset Rac$ — 5, $\forall$ elim
> 7. $Rab \land Rbc$ — 1,2, $\land$ intro
> 8. Rac — 7,6, m.p.

It is accomplished by successive applications of the rule of $\forall$ elim and substitution of the required individual schemata.

A similar situation occurs in trying to show that the inference 'Someone is the student of somebody .'. Someone is the teacher of somebody' is valid. To show this we must first specify that the relation of x being the student of y is the converse of the relation of y being the teacher of x. The rule governing the inference then becomes

11) $\dfrac{\exists x \exists y Rxy}{\forall x \forall y (Rxy \equiv Syx)}$
$\overline{\exists y \exists x Syx}$

11) can be easily shown to be a derived rule, as the reader may verify for himself.

Exercises.

I. Demonstrate by means of the calculus the validity of the following inferences in which only monadic predicates occur. Numbers 8-10 contain both predicates and sentences as recurring constituents.
 1. Numbers 2 and 6-8 of the exercises at the end of Section 21.
 2. Numbers 2, 5, 6, 8, and 10 of Part I of the exercises at the end of Section 23.
 3. Any candidate will be defeated if he fails to use television. Jones will fail to use television. Therefore, if Jones is a candidate he will be defeated.
 4. Jones is a candidate. If anyone is a candidate then someone will be elected to office and serve for four years. Therefore, someone will serve for four years.
 5. Every number is either rational or irrational. All irrational numbers have non-terminating decimal expansions. The expansion of the square root of 4 does terminate. Hence, the square root of 4 is rational.
 6. No candidate will be defeated if he uses television. Some candidate will be defeated. Therefore, some candidate will not use television.
 7. All candidates desire either to serve big business or to serve the public. No candidate worthy of office desires to serve big business. Therefore, if all candidates are worthy of office then all of them desire to serve the public.
 8. Some candidate will be defeated if there is not television time available. If someone is defeated then every contribution will be wasted. There is no television time available. Therefore, if there were contributions then something will be wasted.
 9. Either the argument in the _Phaedo_ is correct or no soul is immortal. But the argument in the _Phaedo_ is mistaken. Therefore, if there are any non-physical entities, none of them that are immortal are souls.
 10. If Descartes is right, then there is no knowledge unless it is certain. All that is certain is known _a priori_. But there is something not known _a priori_. Therefore, either Descartes is mistaken or there is no knowledge.

II. Demonstrate the validity of the following inferences with relational predicates. The premisses of numbers 10-15 must be supplemented by a premiss specifying the type of relation expressed.

1. Smith owns everything that is valuable. This ring is valuable. It follows that Smith must own this ring.
2. John loves Alice. Alice is beautiful. Therefore, John loves someone who is beautiful.
3. Every man loves Raquel Welch. Therefore, if everyone were a man, everyone would love that screen goddess.
4. Every man loves some woman who is beautiful. All bachelors are men. Therefore, all bachelors love some woman.
5. All mice are rodents. Therefore, whatever is larger than a mouse is larger than a rodent.
6. There is a positive integer which all positive integers are equal to or greater than. Therefore, there is some positive integer that is equal to or greater than itself.
7. Anything valuable is costly. Anyone who buys something costly is rich. Someone with a beard bought a valuable diamond. Therefore, someone with a beard is rich.
8. Only those who gamble owe something to every friend. No Protestant gambles. No Protestant owes anything to everybody.
9. Inferences 2, 3, and 5 of Exercises II at the end of Section 24.
10. Professor Smith is Tom's teacher. Therefore, Tom is a student of Professor Smith.
11. Every dog in the country is healthier than every city dog. Therefore, no city dog is healthier than a country dog.
12. Any particular can be characterized by an attribute. Therefore, no attribute is a particular.
13. Some honor student in the class is taller than any boy in the class. Therefore, some honor student is not a boy.
14. Every millionaire keeps at least one mistress. Hector is a millionaire. Hence, there is at least one mistress that is kept by Hector.
15. Dogs are more ferocious than cats. Some Siamese are more ferocious than any rabbit. Therefore, since Siamese are cats, dogs are more ferocious than rabbits.

III. Construct categorical proofs of the following schemata. Use the results of previous proofs in order to introduce derived rules where convenient.

1. $\vdash \forall x (Px \vee -Px)$

2. $\vdash \forall x [-Px \supset (Px \supset Qx)]$

3. $\vdash \exists x Px \vee \exists x Qx \supset \exists x (Px \vee Qx)$

4. $\vdash \forall x (Px \wedge Qx) \supset \forall x Px \wedge \forall x Qx$

5. $\vdash \exists x Px \equiv -\forall x -Px$

6. $\vdash \forall x Px \equiv -\exists x -Px$

7. $\vdash \forall x (Px \supset A) \equiv \forall x Px \supset A$

8. $\vdash \exists x (A \supset Px) \equiv A \supset \exists x Px$

9. $\vdash \forall x \forall y Rxy \equiv \forall y \forall x Rxy$

10. $\vdash \exists x \exists y Rxy \equiv \exists y \exists x Rxy$

11. $\vdash \forall x Rxx \supset \forall x \forall y (Rxy \vee Ryx \supset Rxx)$

12. $\vdash \forall x \forall y (Rxy \vee Sxy) \supset \exists x \forall y Rxy \vee \forall x \exists y Sxy$

13. $\vdash \forall x \forall y \forall z Rxyz \equiv \forall z \forall y \forall x Rxyz$

14. $\vdash \exists x \exists y \forall z Rxyz \supset \forall z \exists x \exists y Rxyz$

29. Predicate Variables and Identity

Predicate Variables. Recall from Section 17 that open sentences such as '___ is red' are represented in our symbolism by open sentence schemata such as Px, with the individual variable x representing the blank position of the subject term. There seems no reason not to represent also open sentences with a subject but a blank in place of a missing predicate. Among these expressions would be 'This book ___', for which predicates such as '...is red', '...is large', or '...is square' could be substituted to form complete sentences. These open sentences could also include those for which a dyadic, triadic, etc. predicate is to be supplied, e.g. 'Tom ___ Mary', 'Bill ___ the book ___ Mary', etc. We shall extend our symbolism to include predicate variables $\alpha, \beta, \gamma, \ldots$ to represent the blanks in these open sentences. Thus, 'This book ___' can be represented by αa, with α standing for the blank position of a predicate and a for 'this book', while 'Tom ___ Mary' can be represented by αab. Predicate schemata $P, Q, R, \ldots$ can be substituted for the predicate variables to represent the filling of the blanks by specific predicates such as '...is red' or '...loves...'. We could substitute P, for example, for α in αa to form Pa and R for α in αab to form Rab. In this manner the variables $\alpha, \beta, \gamma, \ldots$ take as substitution instances predicate schemata representing monadic or relational predicates.

There is an obvious parallel between the use of individual variables $x,y,z,\ldots$ with their substitution instances $a,b,c,\ldots$ and predicate variables $\alpha, \beta, \gamma, \ldots$ with their substitution instances $P,Q,R,\ldots$. There seems no reason not to continue the parallel and besides forming atomic schemata by substitution from the open schemata αa, αab, etc. introduce quantifiers to bind predicate variables and form quantificational schemata. We can thus introduce the quantifiers $\forall \alpha$ and $\exists \alpha$ and form from αa the schemata $\forall \alpha \alpha a$ and $\exists \alpha \alpha a$. Taking a to represent 'this book', the schemata represent respectively 'This book has all attributes' (or 'All attributes hold of this book') and 'This book has some attribute'. 'Some relation holds between John and Mary' would similarly be represented by $\exists \alpha \alpha ab$. Quantification over predicate variables can be combined with quantification over individual variables. We might represent the sentence 'Everything has some attribute', for example, by $\exists \alpha \forall x \alpha x$ and 'Every man bears some relation to some woman' by $\forall x[Px \supset \exists \alpha \exists y(Qy \wedge \alpha xy)]$, with P representing '...is a man' and Q '...is a woman'.

With the introduction of predicate variables and quantifiers to bind them we advance beyond what is called first-order predicate logic, predicate logic restricted to individual variables, to second-order predicate logic. This latter form of logic is needed to represent most of the basic axiomatic systems of mathematics. Since the application of logic to mathematics per se is not within the scope of this book, we shall not consider second-order logic in more detail.

Identity. It is sufficient for our purpose to note that by introducing predicate variables we enable ourselves to define a basic logical term that occurs in all forms of discourse. As we have seen in Sections 16 and 17, 'is' as a form of the verb 'to be' can be regarded as part of the predicate when followed by a general term, e.g. in 'Socrates is wise' and 'Mark Twain is a writer'. This use of the verb is commonly referred to as the "'is' of prediction." When a singular term follows the verb, however, we have a quite different use of 'is'. Here it is not used to ascribe an attribute to the object denoted by the subject, but instead to state the relation of identity. For example, in

     1) Socrates is the wisest teacher of Plato
and 2) Mark Twain is Samuel Clemens

'is' is being used to state the identity of the individuals denoted by different terms. 1) states that the individual denoted by the proper name 'Socrates' is the same as that denoted by the definite description 'the wisest teacher of Plato', while 2) states that the individuals denoted by two different names are the same. The verb 'is' as it occurs in 1) and 2) is referred to as the "'is' of identity."

To represent the 'is' of identity we use the identity sign '='. With a representing the name 'Mark Twain' and b 'Samuel Clemens', the representation of 2) becomes

     3) $a=b$

The identity 2) is true if and only if the individual denoted by the first singular term is the same as that denoted by the second. More generally, any identity of the form $a=b$ is true if and only if a and b denote the same object or have the same extension.

It would be a mistake, however, to think that by stating the truth conditions for $a=b$ in this manner we have provided a definition of the identity sign. Such a definition would be clearly circular, since 'identity' and 'sameness' are synonyms; we would be simply saying that identity holds between a and b when they denote identical objects. An extensional definition of identity in terms of denotation of terms is thus impossible.

A definition is possible, on the other hand, by adopting Leibniz's principle of the "identity of indiscernibles." Somewhat modified this becomes the principle that two terms a and b are related by identity if and only if a necessary and sufficient condition for any attribute being true of the object denoted by a is that it also be true of that denoted by b, that is, if the objects are indistinguishable with respect to every attribute. Using predicate variables Leibniz's principle may be formulated as

$$a=b =_{df} \forall \alpha(\alpha a \equiv \alpha b)$$

It is evident that there is no circularity here, since the material equivalence connective occurs in the definiens, not the identity sign. Notice that this definition requires us to quantify over the predicate variable α. Defined intensionally in terms of attributes holding of denoted objects identity is thus a concept of second-order predicate logic.

The status of identity constitutes a problem central to many areas of philosophy. We have been referring to identity as a type of relation. But this appears to be a mistake if we adopt a restricted conception of a relation.[1] This restricted conception requires that no relational sentence in which a relation is expressed be decomposable into a conjunction of sentences containing monadic predicates. Sentences that can be decomposed in this manner have only the appearance of relational sentences. Consider, for example, the sentences,

    4) John is the same weight as Jim
    5) John is the brother of Jim.

If 4) is true, then there is some particular weight, say 150 pounds, that can be ascribed to John and Jim. 4) can then be reformulated as the conjunction,

    6) John is 150 pounds and Jim is 150 pounds.

In contrast, there seems no way of decomposing 5) into sentences that do not express relations. The relation of being the brother of expressed within 5) is thus a relation proper, in contrast to the "pseudo-relation" of being the same weight as expressed in 4). It is evident that all reflexive relations (cf. Section 24) are pseudo-relations in exactly this sense.

Now consider the identity a=b as defined by $\forall \alpha(\alpha a \equiv \alpha b)$. Let us suppose that there are a restricted number of predicates expressing attributes that can be substituted for the predicate variable. Granted this assumption, it follows that we can express $\forall \alpha(\alpha a \equiv \alpha b)$ as a conjunction of equivalences between sentences expressing attributes shared by a and b. (This would be analogous to our being able to express a quantificational schema of first-order logic such as $\forall x Px$ as a conjunction of atomic schemata if we restrict ourselves to a finite domain (cf. Section 20)). Suppose, for example, that we restrict substitution instances to the predicates '...is red', '...is round', and '...is heavy' represented by P, Q, and R. Then to say 'a is identical with b' is to say 'a is red if and only if b is red and a is round if and only if b is and a is heavy if and only if b is', or a=b if and only if $(Pa \equiv Pb) \wedge (Qa \equiv Qb) \wedge (Ra \equiv Rb)$. More generally, for a finite number of predicates $P_1, P_2, \ldots, P_n$

$$a=b \text{ if and only if } (P_1 a \equiv P_1 b) \wedge (P_2 a \equiv P_2 b) \wedge \ldots \wedge (P_n a \equiv P_n b)$$

This is a decomposition of 'a is identical with b' of the sort characteristic of pseudo-relations. It depends, however, on our being able to specify a finite number of predicates as substitution instances for the predicate variable, a task which may not be realizable.

Identity and Quantity. Leaving aside the question of the status of identity, it is possible to utilize the identity sign as a means of representing the quantity of a general sentence. We have so far been restricting ourselves to the quantifiers 'every' and 'at least one' as indicators of quantity. Now we can represent general sentences in which is indicated the exact numerical quantity of objects in the domain to which the predicates of the sentence are ascribed. The sentence 'There is at least one god', as we have seen, can be represented by $\exists x Px$, with P representing '...is a god'. The sentence 'There is at most one god' can be represented by means of the identity sign by $\forall x [Px \supset \forall y (Py \supset y=x)]$. The sentence 'There is exactly one god' can now be represented by combining these two representations as

$$\exists x [Px \wedge \forall y (Py \supset y=x)]$$

Literally, this reads: 'There is at least one god and if anything else is a god it is identical with that god', or 'There is at least one god and at most one god'. 'There are two gods' would be represented by

$$\exists x \exists y \{ [Px \wedge Py \wedge (x \neq y)] \wedge \forall z [Pz \supset (z=x) \vee (z=y)] \}$$

with $x \neq y$ instead of $-(x=y)$ for 'x is not identical with y'. We state here first that there are at least two things x and y which are P and are distinct ($x \neq y$); and secondly that any thing that is a P is either identical with x or with y, or that there is at most two things which are P. In similar fashion 'There are three gods' would be represented by $\exists x \exists y \exists z \{ [Px \wedge Py \wedge Pz \wedge (x \neq y) \wedge (y \neq z) \wedge (x \neq z) \wedge \forall u [Pu \supset (u=x) \vee (u=y) \vee (u=z)] \}$. It is evident that using the identity sign in this manner enables us to represent any general sentence of the form 'There are n number of Ps'. Conjunctions of non-identities are used to state that the n individuals that are P are distinct, while disjunctions of identities state that no more than these n individuals are P.

It has been claimed by some logician-philosophers that the identity sign also enables us to paraphrase every singular sentence as a general sentence represented by an existential schema. In his "Theory of Descriptions" Russell proposed that a singular sentence whose subject is a definite description can be regarded as equivalent to a complex sentence of two parts, one specifying that there is a unique individual satisfying the description, and the other that the predicate of the original sentence can be ascribed to this individual.[2] For example, the sentence,

      7) The snub-nosed teacher of Plato is wise

is regarded as synonymous with

      8 i) There is exactly one thing that is snub-nosed and a teacher of Plato
      and ii) that thing is wise.

With the predicate schema P for the complex predicate '...is snub-nosed and a teacher of Plato', 8i) becomes represented, as we have seen, by $\exists x[Px \wedge \forall y(Py \supset y=x)]$. With Q for '...is wise', 8ii) becomes Qx, where x represents a pronoun referring back to the thing that is snub-nosed and a teacher of Plato. Sentence 7) as paraphrased by 8) is now represented by

      9) $\exists x\{[Px \wedge \forall y(Py \supset y=x)] \wedge Qx\}$

Where the subject of a singular sentence is a proper name rather than a description we can first substitute for the proper name, Russell claimed, a definite description with the same extension, and then proceed as above. For example, the sentence 'Socrates is wise' would be first paraphrased by substituting for 'Socrates' the definite description 'the snub-nosed teacher of Plato' to form the sentence 7). In this form it can now be represented by 9). In general, for any singular sentence Qa, where a is either a definite description or proper name, Russell claimed there is an equivalent general sentence of the form of 9). In the paraphrase of the singular sentence a category term such as 'thing' replaces the singular subject as the sentence's subject.

Russell used this technique of paraphrase in order to apply logic to questions of ontology and to criticize certain conclusions drawn by the philosopher Meinong from the use of singular terms such as 'Pegasus' or 'Satan' that seem to denote non-existent entities. Russell's technique of paraphrase in order to eliminate singular terms has been since applied by others with a similar end.[3] A serious difficulty has been raised by Strawson, however, concerning the adequacy of such paraphrase in capturing the truth conditions of singular sentences as used in actual communication.[4]

The Predicate Calculus with Identity. The predicate calculus can be extended in order to include rules of inference governing the identity sign. The resulting extension is called the predicate calculus with identity. Our first rule has the effect of stating that it is unconditionally true that an individual is identical with itself.

R15. Identity Introduction: An expression of the form $\nu = \nu$ is categorically provable.

$$\frac{\left|\begin{array}{c} \vdots \\ \psi \end{array}\right.}{\nu = \nu}$$

In other words, from any conclusion of a sub proof a schema of the form $\nu = \nu$ is provable. This rule allows us to enter an identity of the form $\nu = \nu$ at any step of a proof, no matter what the preceding steps may be. The meta-schema ν stands for any free variable and is governed by the principle of substitution of the predicate calculus (cf. Section 26). Within proofs not general sub proofs we must substitute an individual schema for the variable; within general sub proofs, however, substitution is not required.

The second rule needed amounts to a restatement of the definition of identity in terms of Leibniz's principle.

R16. Identity Elimination: From an identity between μ and ν and a schema in which ν occurs, we may infer that schema with μ substituted for ν.[5]

$$\frac{\mu=\nu}{\phi\nu} \\ \phi\mu$$

Again there are two different applications of this rule. Outside of general sub proofs we substitute individual schemata for the variables μ and ν. Here the rule is equivalent to rules such as

$$\frac{a=b}{\phi b}, \quad \frac{b=c}{\phi c}, \quad \frac{c=d}{\phi d}, \quad \ldots$$

Inside general sub proofs the rule is usually applied to variables.

With these rules it is possible to prove that identity is symmetric.

```
1 | a=b    p
2 | a=a    = intro
3 | b=a    1,2, = elim
```

Since step 2 is an identity, it can be introduced regardless of what the preceding step is. Step 3 is derived from 2 by taking a=a as the ϕμ of R16, with ϕ representing '=a', a substituted for the variable represented by μ, and b substituted for ν. Substituting b for the first occurrence of a in step 2 gives us the conclusion b=a. Using this result as a derived rule (the rule of = symmetry), the transitivity of identity is derivable.

```
1 | a=b    p
2 | b=c    p
3 | b=a    1, = sym
4 | c=a    3,2, = elim
5 | a=c    4, = sym
```

These identity rules may be used to show that inferences formulated in an object language are valid. For example, the inference,

> Mark Twain is the author of <u>Huckelberry Finn</u>.
> Mark Twain is either foolish or profound.
> <u>But the author of Huckelberry Finn is not foolish.</u>
> ∴ Mark Twain is profound.

can be shown to be valid by the following conditional proof.

```
1 | a=b      p
2 | Pa v Qa  p
3 | -Pb      p
4 | b=a      1, = comm
5 | -Pa      4,3, = elim
6 | Qb       2,5, d.s.
```

<u>Exercises</u>.

I. Represent the logical form of the following sentences, using the suggested symbolism. Represent the singular sentences 8-10 as general sentences by means of Russell's technique of paraphrase, where necessary by introducing a definite description with the same extension as the proper name in the sentence.

1. There is at least one ruler. (R: '...is a ruler')
2. There is at most one ruler.
3. There is exactly one ruler.
4. Mary loves at least one man. (m: 'Mary'; M: '...is a man'; L: '...loves...')
5. Mary loves at most one man.
6. There are at least two different craters. (C: '...is a crater')
7. There are exactly two craters.
8. The tallest mountain in the world has been scaled. (T: '...is the tallest mountain in the world'; S: '...has been scaled')
9. Satan is evil. (P: the predicate formed from a definite description with the same extension as 'Satan'; E: '...is evil')
10. Satan does not exist.

II. Demonstrate by means of the predicate calculus with identity the validity of the following inferences. Inferences 6-8 require supplying an assumed premiss.
1. If the '68 Democrats' nominee were President, the war would be ended. Humphrey was the '68 Democrats' nominee. The war has not ended. Therefore, Humphrey is not President.
2. Babe Ruth was the Sultan of Swat. Hence, Babe Ruth hit 60 home runs if and only if the Sultan of Swat did also.
3. The heir to the throne waved goodbye. Princess Anne didn't wave goodbye. Hence, Princess Anne must not be the heir to the throne.
4. Prince Charles is the heir to the throne. Prince Charles is the son of Queen Elizabeth. Therefore, the heir to the throne didn't wave goodbye unless the son of Queen Elizabeth did.
5. 3 is the successor of 2. 3 is the square root of 9 and 2 is the first even number. Hence, the square root of 9 is the successor of the first even number.
6. There is at most one President of the United States. Nixon is the President. Hence, Agnew is not.
7. John loves only Alice (or John loves Alice and at most Alice). Therefore, John does not love Mary.
8. There is exactly one person who loves Alice. Alice is the Smiths' eldest daughter. John loves Alice. Hence, the wealthiest lawyer in town does not love the Smiths' eldest daughter.

30. Syntax and Semantics

We are now in a position to discuss the relationship between the natural deduction calculus developed in Chapter III and the present chapter and the decision procedures developed in Chapters II and V. The branch of logic in which calculi are developed for conducting proofs independently of the interpretation of the expressions occurring within them is called the <u>syntax</u> of logic. <u>Semantics</u> is that branch which develops systems of interpretation in which decision procedures are constructed. Whereas the objects of syntax are sentences considered in themselves as forms of expression, the objects of semantics are sentences interpreted as expressing propositions that are either true or false. A third branch of logic is also usually distinguished. This is <u>pragmatics</u>, where sentences are considered as interpreted relative to the purposes of those using them. The distinction in Section 2 between validity and soundness in terms of the different forces with which an inference is used may be regarded as a distinction within pragmatics.[1]

In the present section we shall restrict ourselves to two problems concerning the relationship between syntax and semantics. The first is the problem of priority between these two branches; the second is the technical formal problem of constructing syntactic calculi and determining whether they have the property called "completeness."

<u>The Priority Problem</u>. By a <u>semantic definition</u> of a logical constant we shall mean a <u>definition by means</u> of truth conditions. The truth table definitions of the sentence connectives in Chapter II are clearly semantic in this sense, since we specify in them the truth conditions of a negation, conjunction, etc. in terms of the truth or falsity of their atomic constituents. No such truth table definitions for the universal and existential quantifiers have been given. But we did define the universal quantifier in terms of a conjunction and the existential quantifier in terms of a disjunction for finite domains (cf. Section 20), and truth conditions for conjunction and disjunction can be stated. By a <u>syntactic definition</u> of a logical constant we shall mean any definition not requiring for its statement the interpretation of atomic constituents. We have presented so far two types of definitions in the text. The first is the <u>explicit</u> definitions of sentence connectives of the form $\phi =_{df} \psi$, where the definiendum ϕ contains the connective being defined (cf. Section 7). We defined, for example, the material implication sign in terms of negation and disjunction by the definition $A \supset B =_{df} -A \vee B$. Here we can regard the definiendum $A \supset B$ as an abbreviating expression for the definiens $-A \vee B$. A second type of syntactic definition is the <u>implicit</u> definition of the logical constants by means of the introduction and elimination rules of the natural deduction calculus. An example of such a definition would be that given for the conjunction sign in terms of the rules of conjunction introduction and elimination. The "meaning" of the conjunction sign (and hence of the word 'and' it represents) is specified by the two rules: a conjunction is that from which

either conjunct can be inferred and that which can be inferred if the conjuncts are both given. The implicit syntactic definition of '∧' thus requires only rules of inference applying to molecular sentences of the form of conjunctions, in contrast to the semantic definition requiring interpretation of sentences as expressing true or false propositions.[2]

It would seem that the semantic definitions of the logical constants are presupposed by the choice of primitive rules of the calculus. We choose as primitive only those rules that are valid, those that if followed will never lead us from true premisses to a false conclusion, and the determination of whether or not a rule is valid is made by means of the truth table definitions. The validity of the rules of conjunction introduction and elimination is established, for example, on the basis of the truth table definition of the conjunction sign. This priority of semantic definitions over syntactic would seem to hold for semantics and syntax generally. While we can develop a semantic decision procedure independently of a syntactic calculus, the construction of a syntactic calculus is dependent on a decision procedure. Only by means of one can we justify the selection of primitive rules.

To this it might be objected that the truth table definitions of the connectives themselves presuppose some prior understanding of these connectives. For in stating the truth conditions for molecular schemata in which the connectives occur we are forced to use the same logical words that we are defining. For example, we state that a conjunction A∧B is true if both A *and* B are true, thus using the word 'and' which we are supposedly defining. Again, in defining the material implication sign we state that *if* A is true and B false *then* A⊃B is false, using the expression 'if ... then' which *is* represented by the implication sign. It would seem, therefore, that the semantic definitions are circular, since they use terms that are being defined. Now it seems reasonable to suppose that the prior understanding of such logical words as 'and' and 'if ... then' that is presupposed in the semantic definitions is derived from our use of these words in conducting inferences in accordance with rules. The 'and' used in stating the truth conditions for a conjunction is the 'and' we use in inferring from a conjunction to a conjunct. But it is just such rules that are formulated in our syntactic calculus. Hence this calculus is presupposed by our semantic definitions.

But it is doubtful whether this argument establishes its conclusion. It is true that we can follow rules of inference before being able to justify them (cf. Section 11), and can use a logical word such as 'and' before being able to justify this use. Stating the truth table definition of 'and' would be an instance of this pre-evaluative use. Nevertheless, the general aim of the calculus is not simply that of formulating rules of inference used in reasoning. It is constructed instead as a means of evaluating inferences as valid and establishing that a sentence expresses a tautology. To construct a means for evaluating inferences and to formulate the rules actually used in conducting inferences are clearly distinct. The fact that we can commit errors in deductive reasoning shows that we often use rules of inference for which no justification can be given. The calculus is thus more than a formulation of rules actually in use, and because it requires the validation of its primitive rules it does presuppose the semantic definitions of the logical constants. We could, of course, construct a calculus for some other purpose than evaluating inferences. We might be simply interested in tracing out the consequences of certain arbitrarily chosen primitive rules. But this is not the purpose of the calculus developed in Chapter III and the present chapter, and it is this calculus and its relationship to the semantic decision procedures that we are discussing.[3]

<u>Calculi and Their Properties</u>. Ours has been an informal development of a syntactic calculus. A rigorous formal development would have required, first of all, the specification of those symbols that constitute the basic elements from which schemata are formed. These are called the <u>primitive symbols</u> of the calculus. The primitive symbols of the predicate calculus (without identity) developed here is composed of

    i) sentence schemata: $A,B,C,\ldots$
    ii) sentence connectives: $-, \wedge, \vee, \supset$, and $\equiv$
    iii) grouping indicators: $(,), [,], \{,\}$
    iv) predicate schemata: $P,Q,R,\ldots$
    v) individual schemata and variables: $a,b,c,\ldots$ and $x,y,z,\ldots$
    vi) quantifier symbols: $\forall$ and $\exists$.

Next it would have been necessary to state <u>formation rules</u> by which elements of this vocabulary are combined to form the schemata (or, as they are more commonly called, "well-formed formulae") to which the rules of the calculus are applied. The formation rules for combining the above primitive symbols may be stated as follows:

    i) $A,B,C,\ldots$ are schemata.
    ii) If ϕ is a predicate schema and ν any finite sequence of individual schemata or variables, then $\phi\nu$ is a schema.
    iii) If ϕ and ψ are schemata, then so are $-(\phi)$, $(\phi \wedge \psi)$, $(\phi \vee \psi)$, $(\phi \supset \psi)$, and $(\phi \equiv \psi)$.
    iv) If ϕ is a schema, then $\forall \mu(\phi)$ and $\exists \mu(\phi)$ are schemata, where μ is any variable.

In addition to these rules conventions for eliminating parentheses (the binding conventions and conventions for scope of quantifiers) and replacing them by brackets and braces have been introduced in the text. The formation rules have the effect of preventing sequences of symbols such as ∧A∨Px∀x from being included as schemata, while allowing one such as A∧ ∀x(Px∨Qxy). This latter expression would be formed by applying rule ii) to form from the predicate schemata and variables the open schemata Px and Qxy. To these schemata we would apply rule iii) to form (Px∨Qxy), and then apply rule iv) to prefix the quantifier and form the quantificational schema ∀x((Px∨Qxy)). Finally, we would apply rule iii) again to form from the quantificational schema and the sentence schema A the schema (A∧ ∀x((Px∨Qxy))). Superfluous parentheses could then be eliminated. Only schemata formed in accordance with these formation rules can occur as steps within a proof of the calculus.

Once the primitive symbols have been specified and formation rules stated, two types of calculi can be constructed. A natural deduction calculus such as that developed in Chapter III and the present chapter is constructed by selecting certain primitive rules of inference and syntactic principles. A second type of calculus called an <u>axiomatic calculus</u> can also be constructed. In this form of calculus we select certain schemata called <u>axioms</u> along with rules of inference for deriving from these axioms other schemata called <u>theorems</u>.[4] The axioms selected are tautologies when interpreted and the rules are valid rules that insure that all schemata derived from the axioms are also tautologies. An axiomatic calculus can be constructed which is formally equivalent to the predicate calculus constructed here, that is, in which all schemata for which categorical proofs can be given in the natural deduction calculus can be derived as theorems.[5] By introducing definitions, e.g. by defining '∧', '∨', and '≡' in terms of '-' and '⊃' and '∃x' in terms of '∀x', the number of axioms can be reduced. Similarly, by introducing such definitions in a natural deduction calculus we can reduce the number of primitive rules.

We have seen in Sections 15 and 28 that by virtue of the semantic justification of the primitive rules conclusions of categorical proofs within the natural deduction calculus become tautologies when interpreted. The analogous result for an equivalent axiomatic calculus would be the fact that all theorems provable from axioms by the rules of inference are tautologies, since the axioms are tautologies and the rules valid. We desire also of a calculus that all tautologies be provable, either as conclusions of categorical proofs for a natural deduction calculus or as theorems for an axiomatic calculus. A calculus for which this is possible is said to be <u>complete</u> (or adequate). To prove whether or not various calculi possess the property of completeness is one of the central problems for the branch of mathematics called <u>mathematical logic</u>. During the 1920s and '30s proofs were <u>accomplished</u> for the important types of calculi.

The results of these proofs are somewhat surprising. The sentence calculus has been proved to be complete: if a schema is a tautology as determined by the semantic decision procedure of Chapter II a proof for it within the calculus is in principle possible, though it may still await discovery. The predicate calculus as the extension of the sentence calculus to include rules governing the quantifiers has also been proved to be complete.[6] This is so, despite the fact that it is known that there is a general decision procedure only for monadic predicates (cf. Section 24). Since no procedure exists generally for relational predicates, the best we can say here is that if an expression containing such predicates were determined to be a tautology (even though no method for determining this may in fact be possible) it would be provable within the calculus. The same property of completeness holds of the predicate calculus with identity. But when we extend our calculus to include the quantification over predicate variables in second-order predicate logic it has been proved by Kurt Gödel that the resultant calculus is incomplete, that there are logical truths that cannot be proved, not because a proof has not yet been found, but because a proof is in principle impossible.[7] Since quantification over predicate variables is needed in order to represent the axioms of elementary arithmetic, Gödel's proof has the consequence that no calculus can be constructed in which all truths of arithmetic are provable. A similar consequence follows for most of the other major branches of mathematics. Gödel's proof is of great importance in establishing that proofs of theorems in the major branches of mathematics cannot be "reduced" to proofs within a logical calculus.

It is also of interest to determine whether two other properties hold of a given calculus. The rules of a natural deduction calculus are said to be <u>independent</u> if no one of these rules is derivable from the others. The axioms of an axiomatic calculus have this property of independence if no one of them can be derived as a theorem from the remaining. Definite procedures have been devised for determining whether or not rules or axioms are independent. A set of rules is said to be <u>consistent</u> if no categorical proof of a contradiction can be constructed by means of them. Axioms are consistent if their conjunction fails to express a contradiction, or alternatively, if not every schema can be derived as a theorem from them.

NOTES FOR CHAPTER VI

Section 27.

1. In developing an alternate calculus Quine replaces the rule ∃elim by the rule $\exists\mu\phi\mu \vdash \phi\nu$, with restrictions sufficient to prevent illicit inferences. See his [Methods], Sec. 28 and also Copi [Logic], Ch. 4. The simplicity of this rule is only apparent when balanced by the restrictions required in applying it.

Section 29.

1. That identity is not a relation is argued by Wittgenstein [Tractatus], 5.301ff.

2. See Russell [Denoting].

3. See Ryle [Misleading Expressions] and Quine [Logical View], Essay I for similar applications.

4. See Strawson [Referring].

5. The restrictions should be added that the replaced occurrences of μ neither are nor contain bound occurrences of variables. For a discussion of the need for this restriction see Resnik [Logic], p. 364.

Section 30.

1. The distinction between these branches is due to Morris [Foundations]. In Morris' classification, however, pragmatics is the study of the behavior of language users rather than their purposes in using language.

2. The modern notion of an implicit definition is due to Hilbert. For a discussion of it see the Kneales [Development], pp. 683ff and Pap [Semantics], pp. 202ff.

3. For a discusssion of this issue see Prior [Inference-Ticket] and Belnap [Tonk]. The explanation of the meanings of the logical connectives in terms of their use within formal rules of inference Belnap terms the "synthetic mode" of explanation; the explanation in terms of truth tables is termed by him the "analytic mode." See also Pap [Semantics], pp. 364-366.

4. In an axiomatic calculus the rule of implication introduction of the natural deduction calculus is proven to be a derived rule of inference. The result of the proof is called the <u>deduction theorem</u>. The classic source for the development

of this form of calculus and proofs of its properties is Hilbert and Ackermann [Logic]. See also Church [Logic].

    5. For the nature of the equivalence between the natural deduction calculus developed here and an axiomatic calculus see Leblanc [Deductive Inference], Secs. 1.3, 2.3.

    6. The results of these proofs are conclusions that if ϕ is a tautology, where ϕ is any sentence or quantificational schema, then $\vdash \phi$. These conclusions are among the "meta-theorems" of mathematical logic, theorems about a calculus and not proved within it. For a perspicuous proof of completeness for a set of axioms for the sentence calculus see Basson and O'Connor [Logic], Ch. IV. For a more rigorous proof for the predicate calculus see Henkin [Completeness] or Church [Logic], Secs. 18 and 24. For a general discussion of the basic meta-theorems of predicate logic see the Kneales [Development], Ch. XII. Because of the equivalence between the different forms of calculus, the completeness proofs apply also to the natural deduction calculus developed here.

    7. See Godel [Undecidable Propositions]. For a clear exposition of the methods used by Godel in his proof and an explanation of its consequences for our understanding of the relation between logic and mathematics see Nagel and Newman [Godel's Proof].

    8. For perspicuous proofs of the consistency and independence of a set of axioms for the sentence calculus see Basson and O'Connor [Logic], Ch. IV. For the predicate calculus see Church [Logic], Secs. 41-43.

VII. MODAL LOGIC

31. The Modal Trichotomies

It is a striking fact that in apparently very different areas of discourse we find fundamental trichotomies that exhibit a similar logical structure. Inferences whose validity depends on this structure are evaluated in what is known as modal logic. One such trichotomy we have already encountered in our consideration of predicate logic. There we classified propositions into those that are tautologous, contingent, and contradictory. This classification is exhaustive and exclusive: every proposition must fall into one of these categories and none can belong to more than one. The semantic decision procedures of Chapters II and V enable us to determine in which of these three categories a given proposition belongs. A tautology we can regard as a special kind of proposition whose truth is necessary, while a contradiction is one whose truth is impossible. Our basic semantic trichotomy can be thus stated as that between propositions whose truth is necessary, contingent, and impossible.

This trichotomy can be represented with the help of two logical constants called modal operators that indicate the modal status of a proposition. The sentence 'The proposition expressed by sentence A is necessarily true', or more simply, 'A is necessary', we shall represent by $\Box A$, with $\Box$ as a logical constant called the necessity operator and A a schema standing for the name of some arbitrary sentence. Thus, $\Box A$ may represent 'The proposition expressed by "It is raining or it is not raining" is necessarily true', or more simply in indirect speech, 'It is necessary that either it is raining or it is not'. 'The proposition expressed by the sentence represented by A is possible' will be represented by $\Diamond A$, with $\Diamond$ as the possibility operator. The schema A within the scope of the modal operator is the operand. Note that both $\Box A$ and $\Diamond A$ represent sentences of the meta-language about what is expressed by other sentences. They are not to be confused with a schema such as -A which represents a sentence of an object language (cf. Section 9). Also, unlike -A, $\Box A$ and $\Diamond A$ are not truth functions of their constituent A. The fact that the proposition expressed by A is true does not determine whether or not A is necessary, nor does the fact that it is false determine that A is possible. Necessity and possibility are determined by the kind of proposition A expresses, not by its truth value.

With the necessity and possibility operators we can represent the three basic modalities. That A is contingent (expresses a contingently true proposition) would be represented by $\Diamond A \land \Diamond -A$, while that A is impossible would be represented by $-\Diamond A$. The basic trichotomy of propositions is thus between $\Box A$ (necessity), $\Diamond A \land \Diamond -A$ (contingency), and $-\Diamond A$ (impossibility). The study of the logical relations between necessity and possibility interpreted as indicating the modal status of the truth of a proposition is the domain of what is called <u>alethic modal logic</u>.[1]

This same necessary-contingent-impossible trichotomy has been applied in traditional metaphysics to the description of events and states of affairs. We say of an event that it necessarily occurs when it is an effect that could not but follow some cause, as, for example, boiling of water is necessary when heated to 212 degrees at sea level. Events are also described as contingently occurring when there are no assignable determining causes. A contingent event might be a person's lifting his hand, since given prior events it seems that it might not have occurred. An impossible event is one whose occurrence is excluded, for example, the event of a person jumping unaided 12 feet off the ground.[2] In the same way states of affairs can be described as necessary, contingent, or impossible. It seems necessary that a human being have a heart, but contingent that he have blue eyes, and impossible that he have three arms.

Again, this trichotomy can be represented by expressions containing the necessity and possibility operators $\Box$ and $\Diamond$. Sentences such as

      1) Water necessarily boils when heated to 212 degrees
and 2) A human necessarily has a heart

can be represented by $\Box A$, with A representing in 1) the sentence 'Water boils when heated at 212 degrees' and in 2) 'A human has a heart'. A sentence stating that an event or state of affairs is possible is represented by $\Diamond A$. Again the two operators can be used to represent the modal trichotomy. For any event or state of affairs described by a sentence A we have either $\Box A$, $\Diamond A \land \Diamond -A$ (contingency) and $-\Diamond A$ (impossibility). What we can call <u>physical modal logic</u> is that branch of modal logic studying relations between necessity and possibility interpreted as applying to physical events and states of affairs rather than propositions.[3]

There have been serious questions raised, however, concerning the legitimacy of regarding the physical modalities as having coequal status with the alethic modalities. The basic distinction between the two is that whereas the alethic modalities apply to the truth of propositions and are expressed by sentences in the meta-language, the physical modalities apply to events or states of affairs and are expressed within the object language. In the terminology of the Medieval logicians the physical modalities are among the "<u>de re</u>" modalities, modalities about things, in contrast to the "de dicto" alethic modalities applied to what we say about things. The position of many contemporary philosophers is that all de re modal sentences are but disguised forms of de dicto modal sentences of the meta-language, and when properly understood should be translated into this form. Thus, the de re modal sentence 1) becomes translated into the de dicto meta-linguistic sentence

       3) 'Water boils when heated to 212 degrees' is necessary

where a sentence is mentioned. Similarly, 2) becomes

       4) 'All humans have hearts' is necessary.

The claim is that through such translations the physical modalities are reducible to the alethic, and hence that physical modal logic is not an independent branch of logic.[4] There is less than unanimous agreement with this contemporary view; nevertheless, sufficient doubt has been cast on the status of physical modal logic to justify our not developing this branch further. Modal sentences in which 'necessary', 'contingent', and 'impossible' occur will therefore be interpreted in the following sections as sentences of the meta-language in which object language sentences are mentioned.

    There seem to be three basic modes of belief or knowledge that a person x may have of a proposition expressed by a sentence A. x may be certain (or know) that A expresses a true proposition, he may be uncertain (or ignorant) whether A is true or false, or he may be certain (know) that A is false. The structure of this modal trichotomy is the topic of _epistemic modal logic_. There are a variety of propositions towards which these three types of certainty or knowledge can be directed. One important class are the propositions that constitute the hypotheses of the empirical sciences. At the initial stages of inquiry an investigator x may be uncertain whether a hypothesis H is true or false. If H is falsified by an experiment (and he is certain of auxiliary assumptions), then x would be certain that H is false. If H survives a certain number of experimental tests (with appropriate controls), x may be certain that H is true. Another important class of propositions to be considered are those expressed by formulas within mathematical theories. If a mathematician x can prove that a certain formula F can be derived from the axioms of a theory, then x can be certain that F is true relative to these axioms. If a proof can be provided for the negation of F, then x is certain that F is false in the same manner. But if neither F nor its negation can be proved (as is the case for Fermat's Last Theorem), then x would be uncertain about F.

This epistemic trichotomy can also be represented by operators for certainty and for uncertainty of falsity (or unfalsified). Within the modal logic that results there are important similarities to alethic modal logic. Two important divergences should be noted, however: it does not follow that if x is certain of A that A is true (though if x knows that A, then A is true), whereas if A is necessary, then A is true in alethic logic; and secondly, if A is true, then it does not follow that a person is uncertain whether A is false (he may be ignorant of its truth), whereas if A is true it follows that it is also possible. The epistemic modalities are on much more secure footing than are the physical. We shall also bypass this branch in the sections that follow, however, in order to concentrate on two branches that have been more fully developed.[5]

The reader will have noticed that two of the trichotomies outlined so far correspond to the contrasts drawn by those seeking an alternative to 2-valued logic (cf. Section 25). The uncertainty which we may have as to whether it will rain tomorrow contrasts, as we saw, with the certainty of the truth of the proposition and the certainty of its falsity. The argument from undecidability thus sets forth an epistemic modal trichotomy rather than a contrast between three types of truth values. Similarly, the argument from indeterminacy sets forth the modal contrast between a contingent event such as a person's driving his car tomorrow and one that is necessary or impossible.

The final trichotomy we shall consider is that between an action being obligatory, its being indifferent (neither it nor its negation being obligatory), and its being forbidden. For example, a person might be (legally) obligated to pay his taxes, while it is indifferent whether he files his return in January and forbidden for him to deduct vacation travel as a business expense. The logical relations between the terms of this trichotomy is the subject matter of <u>deontic modal logic</u>. We shall consider this branch in some detail in the next chapter.

The contrast between the different branches of modal logic can be conveniently summarized in a table listing the different trichotomies and the type of elements to which the modalities are applied.

Branch of modal logic	Trichotomy	Applications
alethic	necessary-contingent-impossible	truth of proposition
physical	necessary-contingent-impossible	events, states of affairs

Branch of modal logic	Trichotomy	Applications
epistemic	certain-uncertain-certain not	truth of propositions as believed by a person
deontic	obligatory-indifferent-forbidden	actions

We shall devote our attention in the remainder of this chapter to alethic modal logic. In the next section a contrast will be drawn between two different senses in which propositions can be said to be necessary, contingent, or impossible, and in Sections 33 and 34 a semantic decision procedure and calculus will be developed for this branch of logic.

32. Interpretations of the Alethic Modalities

The formal structure of modal logic is very simple and parallels closely that of the logic with which we are already familiar. The interest in this form of logic is perhaps chiefly due to the philosophic importance of the various interpretations we can give to the modal trichotomies. We proceed now to introduce some terminology used in the interpretation of the alethic modalities, and to briefly discuss some of the philosophic issues that can be raised within this branch of logic.

Modal Relations. There are two basic modal relations holding between propositions expressed by sentences A and B. The first of these is the relation of entailment of B by A which holds if and only if it is necessary that if A is true B is also. This relation we symbolized in Section 9 by $A \Rightarrow B$, where it is understood that A and B stand for names of sentences in the object language and not the sentences themselves. If A represents the conjunction of premises of a valid inference and B its conclusion, then we would have the entailment relation $A \Rightarrow B$. Entailment can be defined in terms of either necessity or possibility as follows:

$$\text{i) } A \Rightarrow B =_{df} \Box(A \supset B) =_{df} -\Diamond(A \land -B)$$

A second basic modal relation is that holding between two propositions when they are compatible (or consistent). Propositions expressed by sentences A and B are said to be compatible if and only if it is possible for A and B to both be true. Symbolizing the compatibility relation by $A \circ B$, it may be defined in terms of either possibility or entailment.

$$\text{ii) } A \circ B =_{df} \Diamond(A \land B) =_{df} -(A \Rightarrow -B)$$

Two propositions expressed by A and B are said to be <u>incompatible</u> (inconsistent) when A∘B is false, i.e. when $-◊(A∧B)$ or $A ⇒ -B$. Logically contradictory propositions, e.g. that 2+2=4 and that 2+2≠4, are obvious incompatible, for the truth of one would entail the falsity of the other.[1]

A proposition is said to be <u>conditionally necessary</u> when it is necessarily true only under the assumption of the truth of some other proposition. Thus, the proposition expressed by B is conditionally necessary if and only if there is some proposition A such that $A ⇒ B$. The conclusions of almost all valid inferences would be conditionally necessary, for they are necessarily true only if the inferences' premises are true. The propositions expressed by empirical laws such as Galileo's law of acceleration are conditionally necessary relative to theories such as Newton's theory of gravitation from which they can be deduced. A <u>categorically necessary</u> proposition, is, in contrast, a proposition whose necessity does not depend on the truth of some other proposition. Such would be all logically necessary propositions such as the tautologies of predicate logic. Since the primitive rules of the predicate calculus can be given a semantic justification, the calculus provides a means of establishing both conditional and categorical necessity. The conclusion of a conditional proof is conditionally necessary relative to the premises from which it is derived, while the conclusion of a categorical proof is categorically necessary.[2]

The conditional-categorical distinction holds also for the modalities of contingency and impossibility. A proposition expressed by B is said to be conditionally contingent relative to A if neither $A ⇒ B$ nor $A ⇒ -B$, or if both B and its negation are compatible with A, that is, if $◊(A∧B)$ and $◊(A∧-B)$. Let B express the proposition that a person performs a certain act voluntarily and A express the proposition in which relevant laws of nature and antecedent conditions are expressed. Then B is contingent relative to A, for if $A ⇒ B$ the action would be determined (and hence not voluntary), while if $A ⇒ -B$ the action could not have been performed. B is said to be conditionally impossible relative to A if $A ⇒ -B$. Categorically contingent and impossible propositions are those whose modal status does not depend on the truth of some other proposition, e.g. logically contingent and contradictory propositions. If a proposition is conditionally contingent, it follows that it is categorically contingent, but the latter does not entail the former. If a proposition is categorically impossible, e.g. a logical contradiction, it is impossible relative to any other proposition, but again the converse does not hold.

<u>Two Types of Analyticity</u>. An <u>analytic sentence</u> is a sentence which can be judged true on the basis of only information about the meanings of its constituent terms. These are contrasted with

synthetic sentences whose evaluation requires information additional to that about meaning. There are usually distinguished two types of synthetic sentences: *synthetic a posteriori sentences* whose truth or falsity is determined by means of empirical evidence and *synthetic a priori sentences* for which no empirical test is available. Two types of analytic sentences can be distinguished. One is constituted by those whose truth is dependent on definitions we give to constituent logical words such as 'not', 'and', and 'every', but is independent of the meanings of constituent subject and predicate terms in the sense that any other terms could be substituted for them within the sentence and the sentence remain true. Examples of these would be

1) John is either tired or not tired

whose truth depends on the meanings of 'or' and 'not' as defined in truth tables but is independent of 'John' and 'tired', and

2) Everything is such that if it is red then it is red

whose truth depends on the meaning of 'every' and 'if...then' but is independent of the predicate 'red'. Sentences such as 1) and 2) we shall call *formally analytic* sentences.

We shall call sentences *materially analytic* when their truth does depend on the meanings of constituent subjects and predicates. One such sentence would be

3) All bachelors are unmarried

since it is by virtue of its subject term being synonymous with 'unmarried man' that it expresses a necessarily true proposition. It is impossible to judge the proposition false, since the predicate expresses a criterion used to identify what is referred to by the subject. Another such sentence would be

4) If this book is red then it is colored

since the sentence's truth derives from the meanings of the predicates '...is red' and '...is colored'. From the logical form alone of 3) and 4) it is impossible to determine their truth, as is the case for 1) and 2). Instead, their truth depends on the semantic content of their non-logical terms.

A parallel can be drawn between the formal-material distinction for analytic sentences and the categorical-conditional distinction for necessary propositions. Formally analytic sentences express categorically necessary propositions of the logical kind, the tautologies of predicate logic. Materially analytic sentences such as 3) and 4), on the other hand, can be regarded as conditionally necessary propositions if we assume what we shall call *meaning-specifying sentences*, sentences specifying the meaning of the subject and predicate terms on which the truth of the analytic sentences depend.[3] For example, 'All bachelors are unmarried' is necessary relative to the meaning-specifying sentence 'All bachelors are unmarried men', since the latter entails it. Similarly, 4) can be regarded as necessary relative to 'Everything red is colored'. If B is taken as representing some materially analytic sentence, then we can interpret the modal expression $\Box B$ as an ellipsis for $A \Rightarrow B$, with A representing some implied meaning-specifying sentence. Thus to say that 3) expresses a necessary proposition is to say, in effect, that there is some implied meaning-specifying sentence, namely 'All bachelors are unmarried men', which expresses a proposition that entails it. Materially analytic sentences would seem to also include the theorems that can be derived within the various systems of mathematics. Here the meaning-specifying sentence would be the conjunction of the axioms of the system. For example, the sentence 'The angles of a triange equal 180 degrees' can be regarded as expressing a necessary proposition relative to the proposition expressed by the conjunction of the axioms of Euclidean geometry. Relative to the axioms of a non-Euclidean geometry the proposition would be necessarily false.

It might be asked whether meaning-specifying sentences such as 'All bachelors are unmarried men' are materially or formally analytic. If they are the former, then some other meaning-specifying sentence A' could be introduced, which in turn would require another sentence A'' entailing it to be materially analytic, which in turn would require A''', and so on indefinitely. But this process must finally terminate. There must, therefore, be meaning-specifying sentences that are not themselves materially analytic. At the same time, however, these sentences are not true by virtue of logical form and hence not formally analytic. There have been two alternative ways of classifying them that have been seriously proposed.

1. We can regard them as disguised synthetic sentences providing a description of how persons within the linguistic community use the constituent words of the materially analytic sentence. Thus, 'All bachlors are unmarried men' can be regarded as a disguised form of the sentence 'Within the English-speaking community "bachelor" is used in a sense synonymous with "unmarried man"'. If this usage were to change and 'bachelor' used in a sense compatible with 'married man', then 3) would cease to be materially analytic.

2. We can regard meaning-specifying sentences as announcements of decisions on our part to use the constituent words in certain ways. As such announcements they would not express true or false propositions. Under this interpretation 'Everything red is colored' is regarded as equivalent to the

announcement 'We agree to use "red" as a color word'. These decisions can be changed, and the changes would again affect the status of sentences like 3) and 4).

Our decision as to which alternative to accept will not affect the central point that all analytic sentences, whether formal or material, express necessary propositions. Those sentences that are formally analytic express categorically necessary propositions; those materially analytic can be regarded as expressing propositions whose necessity is conditional on some specification of the meaning of its subject and predicate terms. In contrast, a synthetic a posteriori sentence whose truth depends on empirical evidence will invariably express a contingent proposition, a proposition whose falsity is possible.

Extension to Inferences. The formal-material distinction can be easily extended to inferences. A *formally valid inference* is one valid by virtue of logical form; its premisses logically entail its conclusion. A *materially valid inference*, on the other hand, is valid by virtue of the meaning of one or more of its constituent subject and predicate terms. Such would be

5) This book is red.
∴ This book is colored.

whose validity depends on the meanings of 'red' and 'colored'. We can regard the conclusion of an inference like 5) being entailed by its premiss relative to an assumed meaning-specifying premiss. By making explicit this premiss we have the formally valid inference,

Everything red is colored.
This book is red.
∴ This book is colored.

In general, any materially valid inference of the form B ∴ C can be changed to a formally valid inference A ∧ B ∴ C if A specifies the meanings of the constituents of B and C upon which the entailment B ⇒ C depends.

This formal-material distinction can also be applied to analytic sentences and valid inferences containing relational predicates. The sentence 'If Professor Smith is the teacher of John, then John is a student of Professor Smith' can be regarded as materially analytic, since its necessity is relative to the proposition that the relation of being a student is the converse of the teaching relation (cf. Section 24). Similarly, the inference,

John is married to Alice.
∴ Alice is married to John.

can be regarded as materially valid, its premiss entailing the conclusion only on the condition that we assume a premiss specifying that the relation of being married is symmetrical. Adding this premiss as we did in Section 24 we have the formally valid inference,

> If anyone is married to another then that other is married to him.
> John is married to Mary.
> ∴ Mary is married to John.

<u>The Problem of the Synthetic A Priori</u>. As mentioned above, synthetic sentences have been traditionally classified into those a posteriori, i.e. those empirically testable, and those a priori whose truth or falsity is not ascertainable by empirical procedures. Synthetic a posteriori sentences express logically contingent propositions, since we must always admit the possibility of their being falsified by some empirical test. The same possibility does not seem open for the propositions expressed by what some philosophers have regarded as synthetic a priori sentences, e.g. 'No object can be both red and green all over at the same time' or the philosophic conclusions 'Persons exist' and 'Propositions can only be either true or false'. Synthetic a priori sentences thus seem to express necessary propositions which if true cannot possibly be false. Now these propositions do not seem conditionally necessary. Their truth does not depend on the truth of some meaning-specifying sentence, for then they would be analytic. We are thus lead to say that synthetic a priori sentences express a kind of categorical necessity that is not the logical necessity of a tautology.

A number of philosophers have maintained, however, that only analytic sentences express necessary propositions and that there is only one type of synthetic sentence, that of the a posteriori variety. What have been taken to be synthetic a priori sentences, they argue, can be shown after careful analysis to be either analytic or disguised announcements of decisions on our part to use terms in certain ways. Under this view categorical necessity is restricted to logical necessity as expressed by formally analytic sentences.[4]

The issues involved in this dispute are complex. It is sufficient for our purposes to leave open the possibility of necessity being interpreted in the categorical non-logical sense in the way advocated by the defenders of the synthetic a priori. Accordingly, the symbolic expression '$\Box A$' that we use for alethic modal logic in the next two sections will be regarded as capable of being interpreted in terms of the three types of necessity outlined in this section: logical necessity as expressed by formally analytic sentences, categorical non-logical necessity as expressed by synthetic a priori sentences, and the conditional necessity expressed by materially analytic sentences. It can thus be used to represent any of the three following modal sentences:

'It is raining or not raining' is necessary (categorical-logical)

'Every proposition true or false' is necessary (categorical-non-logical)

'All bachelors are unmarried' is necessary (conditional on meaning specification)

A similar variety of interpretations can be given for the other basic expression '$\Diamond A$'. This can be interpreted in terms of logical, categorical but non-logical, or conditional possibility. The logical relations that are shown to hold between the various modalities in the next two sections are not affected by these differences of interpretation.

Exercises. Determine for each of the following sentences: 1) whether it is analytic or synthetic; 2) if analytic, whether it is formally or materially analytic; and 3) if materially analytic, what meaning-specifying sentence can be supplied in order to regard it as expressing a conditionally necessary proposition.
1. All swans are white or not white.
2. All swans are white.
3. All swans are birds.
4. All vixens are foxes.
5. John is older than Alice.
6. If John is older than Alice, then Alice is not older than John.
7. No integer is both positive and negative.
8. All squares of integers are positive.

33. Semantic Decision Procedure

The decision procedure for alethic modal logic enables the evaluation of modal inferences as valid or invalid and of modal sentences as to whether or not they express logical truths.[1] This procedure turns out to be analogous to the decision procedure for monadic predicate logic. This is to be expected, since the structures of modal and monadic predicate logic are the same. Our being able to define the necessity and possibility operators in terms of each other suggests this. For A to be necessary is for it to be impossible that the negation of A be true, while for A to be possible is for it to be not the case that its negation is necessary. We have accordingly, the two following definitions:

i) $\Box A =_{df} -\Diamond -A$
ii) $\Diamond A =_{df} -\Box -A$

These can be seen to correspond to the definitions of the universal quantifier in terms of the existential and the existential in terms of the universal.

In like manner all tautologies of monadic predicate logic can be seen to have as analogues logical truths of modal logic, with $\forall x$ and $\exists x$ correlated to $\Box$ and $\Diamond$ and atomic open schemata $Px, Qx, Rx, \ldots$ within the scope of the quantifiers correlated to sentence schemata $A, B, C, \ldots$ within the scope of the modal operators. For example, corresponding to the logical equivalence between $\forall x(Px \wedge Qx)$ and $\forall xPx \wedge \forall xQx$ in predicate logic is the logical equivalence between $\Box(A \wedge B)$ and $\Box A \wedge \Box B$. This structural similarity can perhaps be made intuitively plausible by noting that in the logical sense of necessity $\Box A$ is interpreted as saying that A is true under all possible interpretations of the constituents of A and $\Diamond A$ as saying that A is true under some possible interpretations. Hence the parallel between the structures of 'necessity' and 'possibility' and 'all' and 'some'.

$\Diamond$-Constituents and Their Interpretation. It seems evident that every proposition expressed by a sentence A must have attributed to it one of three modalities: A must be either necessary, contingent, or impossible. These possible modal statuses of A are listed in the following truth table:

$\Diamond A$	$\Diamond -A$	
T	T	(contingency)
T	F	(necessity)
F	T	(impossibility)
F	F	(meaningless)

As indicated to the right, under the first interpretation A is contingent, since both A or its negation is said to be possible. Under the second interpretation A is necessary, since A's negation is denied to be possible. The third interpretation would have A impossible, since it is denied that A is possible. The fourth would have neither A nor its negation possible. Since this would require what is plainly absurd, namely that a proposition be both impossible and necessary, we must exclude this last possibility. The three remaining interpretations constitute a trichotomy that besides being exhaustive is exclusive: every proposition must have one modal status but no more than one.

We shall call modal schemata such as $\Diamond A$ and $\Diamond -A$ possibility-constituents or $\Diamond$-constituents. A single sentence A generates two $\Diamond$-constituents whose three possible interpretations state the possible modal statuses of the proposition A expresses.

For two schemata A and B we can generate $2^2=4$ ◊-constituents. Each of these will be a modal schema in which the possibility operator ◊ operates on a conjunction of each of the two schemata or its negation. As before for predicate logic (cf. Section 22) we shall abbreviate conjunctions of sentence schemata as operands within ◊-constituents by juxtaposition and negation of such sentence schemata by writing a bar over the negated schema. The four ◊-constituents and their $2^4=16$ possible interpretations can be listed in the following truth table:

◊(AB)	◊(A$\bar{B}$)	◊($\bar{A}$B)	◊($\bar{A}\bar{B}$)
T	T	T	T
T	T	T	F
T	T	F	T
T	T	F	F
T	F	T	T
T	F	T	F
T	F	F	T
T	F	F	F
F	T	T	T
F	T	T	F
F	T	F	T
F	T	F	F
F	F	T	T
F	F	T	F
F	F	F	T
~~F~~	~~F~~	~~F~~	~~F~~

If we again exclude the last interpretation (this is indicated by drawing a line through it), 15 possible interpretations remain. In general for n constituent schemata we can generate 2^n ◊-constituents of which there are $2^{2^n}-1$ possible interpretations. A decision procedure for alethic modal logic is enabled by the fact that every modal schema with n constituent sentence schemata can be expanded into a distributive normal form in which it becomes a truth function of some of its 2^n possible ◊-constituents.

Evaluating for Validity and Logical Truth. The expansion of modal schemata into their distributive normal form parallels that for predicate logic. The expansion is enabled by three logical facts: 1) the necessity operator can be defined in terms of the possibility operator; 2) every sentence schema can be expanded into a logically equivalent disjunctive normal form; and 3) the possibility operator ◊ can be distributed over the disjunction sign. 3) can be justified by the intuitive obviousness of the fact that if it is possible for the weather to be either fair or cloudy, then it is possible for it to be fair or it is possible for it to be cloudy. And in general, for two sentences A and B if ◊(A∨B) then ◊A ∨ ◊B. 1) has been established by definition i) of the necessity operator at the beginning of this section; 2) has been established in Section 10.

The expansion to distributive normal form proceeds by successively applying to a given modal schema the three logical facts. As an illustration we expand the schema $\square(A \wedge B)$.

$$\square(AB)$$
$$- \Diamond \overline{(AB)} \qquad \text{(definition of } \square \text{)}$$
$$- \Diamond (\overline{AB} \vee \overline{A}B \vee A\overline{B}) \qquad \text{(replacing } \overline{(AB)} \text{ by its disjunctive normal form)}$$
$$- [\Diamond(\overline{AB}) \vee \Diamond(\overline{A}B) \vee \Diamond(A\overline{B})] \qquad \text{(distribution of } \Diamond \text{ over disjuncts)}$$

The final step in the expansion is a truth function of $\Diamond$-constituents, and is hence in distributive normal form.

To evaluate a modal inference as valid or invalid we simply represent its logical form, expand its premisses and conclusion into distributive normal form, and determine from a truth table listing the possible interpretations of $\Diamond$-constituents whether it is possible for the premisses to be true and the conclusion false. As an example we evaluate the inference,

1) It is possible that it is raining and possible that it is cloudy.
∴ It is possible that it is both raining and cloudy.

The logical form of 1) is

2) $$\frac{\Diamond A \wedge \Diamond B}{\Diamond (A \wedge B)}$$

The distributive normal form of the premiss is arrived at through the following expansion:

$$\Diamond A \wedge \Diamond B$$
$$\Diamond(AB \vee A\overline{B}) \wedge \Diamond(AB \vee \overline{A}B)$$
$$[\Diamond(AB) \vee \Diamond(A\overline{B})] \wedge [\Diamond(AB) \vee \Diamond(\overline{A}B)]$$

The conclusion is already in normal form. Since there are but three $\Diamond$-constituents in the premisses and conclusion of 2), we can write an abbreviated truth table with altogether eight possible interpretations.

① $\Diamond(AB)$	② $\Diamond(A\overline{B})$	③ $\Diamond(\overline{A}B)$	(① ∨ ②) ∧ (① ∨ ③)	①
T	T	T	T	T	T	T
T	T	F	T	T	T	T
T	F	T	T	T	T	T
T	F	F	T	T	T	T
F	T	T	T	T	T	F
F	T	F	T	F	F	F
F	F	T	F	F	T	F
F	F	F	F	F	F	F

As for predicate logic, we label ◊-constituents and express the premisses and conclusion as truth functions of the labels. As can be seen in the fifth interpretation of the truth table, inference 1) is invalid. Taking A to represent 'This book is red' and B to represent 'This book is green', we would have a true premiss and false conclusion.

The evaluation of the logical status of propositions expressed by modal sentences proceeds in similar fashion. As an example we evaluate a sentence of the form

$$3) \quad \Box(A \land B) \equiv \Box A \land \Box B$$

The distributive normal form is gained in the following steps:

$$\Box(AB) \equiv \Box A \land \Box B$$
$$-\Diamond\overline{(AB)} \equiv -\Diamond\overline{A} \land -\Diamond\overline{B}$$
$$-\Diamond(\overline{A}B \lor \overline{AB} \lor A\overline{B}) \equiv -\Diamond(\overline{A}B \lor \overline{AB}) \land -\Diamond(A\overline{B} \lor \overline{AB})$$
$$-[\Diamond(\overline{A}B) \lor \Diamond(\overline{AB}) \lor \Diamond(A\overline{B})] \equiv -[\Diamond(\overline{A}B) \lor \Diamond(\overline{AB})] \land -[\Diamond(A\overline{B}) \lor \Diamond(\overline{AB})]$$

With three ◊-constituents the truth table has the following form:

① ◊(A̅B)	② ◊(A̅B̅)	③ ◊(AB̅)	-[① ∨ (② ∨ ③)]	≡	-(① ∨ ②) ∧ -(③ ∨ ②)
T	T	T	F T T	T	F T F F T
T	T	F	F T T	T	F T F F T
T	F	T	F T F	T	F T F F T
T	F	F	F T F	T	F T F T F
F	T	T	F T T	T	F T F F T
F	T	F	F T T	T	F T F F T
F	F	T	F T T	T	T F F F T
F	F	F	T F F	T	T F T T F

The truth table shows 3) to be a logical truth, and hence that $\Box(A \land B) \Leftrightarrow \Box A \land \Box B$. Interpreting necessity in its logical sense this states that the conjunction of A and B is a tautology if and only if both A and B are tautologies.

As can easily be verified by the reader, the following logical relations also hold in alethic modal logic:

$$4) \quad \Box A \lor \Box B \Rightarrow \Box(A \lor B)$$
$$5) \quad \Diamond(A \lor B) \Leftrightarrow \Diamond A \lor \Diamond B$$
$$6) \quad \Diamond(A \land B) \Rightarrow \Diamond A \land \Diamond B$$
$$7) \quad \Box(A \supset B) \Rightarrow \Box A \supset \Box B$$

Interpreting necessity again in its logical sense, 7) states that if A entails B, then if A is a tautology so is B. The parallel with the basic logical relations in monadic predicate logic listed in Section 23 should be obvious.

It can also be verified that the following two logical relations also hold in this logic:

8) $\quad -\Diamond A \Rightarrow \Box(A \supset B)$
9) $\quad \Box B \Rightarrow \Box(A \supset B)$

8) states that if a proposition is impossible then it entails any other proposition. Hence, '2+2=4 and 2+2≠4' entails 'The moon is made of green cheese', since that 2+2=4 and 2+2≠4 is a contradiction. 9) states that a necessary proposition is entailed by any other proposition, and hence 'The moon is made of green cheese' entails 'Either 2+2=4 or 2+2≠4'. These counter-intuitive results have been termed the "paradoxes of entailment" and have been compared with the "paradoxes of material implication" discussed in Section 7. The results are paradoxical, however, only if the entailment relation is taken in some other sense than the relation between the premisses and conclusion of a valid inference as determined by a truth table, since the inference from A and -A to B is clearly valid, no matter what A and B are taken to represent. To avoid the "paradoxes" it has been proposed that entailment be defined in such a way that A ⇒ B only if A is in some sense "relevant" to B. Whether or not there is a need for this restriction is, however, controversial.[2]

Modal and Non-Modal Sentences as Recurring Constituents. It is possible to extend the procedure just developed to the evaluation of inferences and sentences whose validity or logical status depends on the recurrence of both modal sentences and sentences that are not modal. Consider, for example, the inference,

10) Either the angles of a triangle equal 180 degrees or it is necessary that parallel lines intersect or diverge.
But it is impossible for parallel lines to diverge.
∴ The angles of a triangle equal 180 degrees.

The constituents of this inference are the non-modal sentence 'The angles of a triangle equal 180 degrees' and the modal sentences 'It is necessary that parallel lines intersect or diverge' and 'It is impossible for parallel lines to diverge'. The form of 10) is thus

11) $\quad A \vee \Box(B \vee C)$
$\quad\quad \underline{-\Diamond C}$
$\quad\quad A$

with the sentence schema A for the non-modal sentence. With its premisses and conclusion in distributive normal form, 11) becomes

12) $\quad A \vee - \Diamond(\overline{BC})$
$\quad\quad\underline{-[\,\Diamond(BC) \vee \Diamond(\overline{BC})\,]}$
$\quad\quad A$

Since there is one non-modal constituent and three $\Diamond$-constituents, the truth table for evaluating 12) will have altogether 16 possible interpretations.

A	① $\Diamond(\overline{BC})$	② $\Diamond(BC)$	③ $\Diamond(\overline{BC})$	$A \vee -$①	$-[$ ② $\vee$ ③ $]$	A
T	T	T	T	T	F	T	T
T	T	T	F	T	F	T	T
T	T	F	T	T	F	T	T
T	T	F	F	T	T	F	T
T	F	T	T	T	F	T	T
T	F	T	F	T	F	T	T
T	F	F	T	T	F	T	T
T	F	F	F	T	T	F	T
F	T	T	T	F	F	T	F
F	T	T	F	F	F	T	F
F	T	F	T	F	F	T	F
F	T	F	F	F	T	F	F
F	F	T	T	T	F	T	F
F	F	T	F	T	F	T	F
F	F	F	T	T	F	T	F
F	F	F	F	T	T	F	F

The truth table shows 10) to be an invalid inference, since in the last interpretation the premisses are both true, but the conclusion is false.

Any modal inference such as 8) can be evaluated in similar fashion. The columns in the truth tables used in the evaluation will be headed like the one above by non-modal schemata as well as $\Diamond$-constituents.

<u>Exercises</u>. Evaluate the following modal inferences as valid or invalid by means of the method of distributive normal forms. Note that inferences 9 and 10 include non-modal sentences among their constituents.
   1. It is contingent that the earth has one moon. Therefore, it is contingent that the earth does not have one moon.
   2. There may be life on Mars. That there is life on Mars entails that there is oxygen in its atmosphere. Hence, there may be oxygen in Mars' atmosphere.
   3. That there is life on Mars entails there is oxygen in its atmosphere. Hence, it is possible that there is life on Mars and oxygen in its atmosphere.
   4. It is necessary that if animals can communicate they have a language. It is possible that they can communicate. Therefore, it is possible that animals can communicate and have a language.

5. It is necessary that either the universe is infinite or that it has a boundary. It follows therefore that the universe must be infinite or it must have a boundary.

6. It is impossible for this book to be red and not colored. It is contingent that this book is colored. Hence, it is contingent that it is red.

7. That 7 is less than 9 entails that 9 is greater than 7. That 7 is an odd number greater than 6 is compatible with its being less than 9. It follows therefore that it is possible both for 7 to be an odd number greater than 6 and for 9 to be greater than 7.

8. That a triangle has three sides does not entail that its angles equal 180 degrees. That its angles equal 180 degrees is incompatible with its being a non-Euclidean triangle. Hence, for a triangle to have three sides does not entail its being non-Euclidean.

9. If there is life on Mars there must be both moisture and an atmosphere. But it is impossible for there to be moisture there Hence, there is no life on Mars.

10. That this figure be a triangle and its angles equal to 180 degrees are incompatible unless the triangle is Euclidean. It is possible for this figure to have angles equalling 180 degrees. Hence, the triangle is Euclidean.

34. The Alethic Modal Calculus

The alethic modal calculus is an extension of the predicate calculus (including identity) in which rules governing the necessity and possibility operator are added.[1]

Rules Governing the Necessity Operator. In light of the analogy between the decision procedures for predicate and modal logic it is not surprising that the rules for the modal operators are analogous to those for the universal and existential quantifiers. The analogy should be apparent for the introduction and elimination rules for the necessity operator.

R17. *Necessity Introduction: From a proof of a schema* ϕ *within a necessity sub proof we may infer the modal schema* $\Box \phi$.

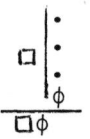

$$\frac{\Box \begin{vmatrix} \cdot \\ \cdot \\ \cdot \\ \phi \end{vmatrix}}{\Box \phi}$$

The following inference could be justified by this rule:

```
  .  |   |  .
  .  | □ |  .
  .  |   |  .
n-1  |   | A∧B⊃C
n    | □(A∧B⊃C)    n-1, □ intro
```

That a sub proof is a necessity sub proof is indicated by the necessity operator to its left. Reiteration into such a sub proof is restricted in such a way as to insure that the rule of necessity introduction will never lead from true premises to a false conclusion. The restrictions on reiteration are established by the following

> **Principle of Modal Reiteration:** <u>Only schemata prefixed by the necessity operator can be reiterated into necessity sub proofs, and after reiteration the operator must be dropped.</u>

The following step would be in accordance with the principle:

```
1 | □(A⊃ B)        p
2 | □ | A⊃ B       1,r
```

If there were no necessity operator prefixed to the schema A⊃B in step 1, step 2 would be forbidden. The effect of this restriction is to prevent inferences such as

```
1 | A⊃ B           p
2 | □ | A⊃ B       1,r (forbidden)
3 | □(A⊃ B)        2, □ intro
```

From the fact that A⊃B is true it certainly does not follow that A⊃B is necessarily true. The requirement that □ be dropped upon being reiterated into necessity sub proofs has the effect of preventing the inference from □A to □□A. The question of whether or not such an inference should be prevented will be raised shortly.

The following proof illustrates the use of the necessity introduction rule and the principle of modal reiteration.

```
1 | □(A⊃ B)        p
2 | □ | A⊃ B       1,r
3 |   | -B⊃ -A     2, contrapos
4 | □(-B⊃ -A)      2-3, □ intro
```

Note the parallel between this proof and proof 1) of Section 26 involving the universal quantifier rules. The sole difference in the form of the proof is due to the fact that the principle of modal reiteration requires us to drop the operator when reiterating into a necessity sub proof, whereas in using the quantifier reiteration principle we retain the quantifier and then eliminate it once inside a general sub proof. The strategy employed in a proof such as this is analogous to the universal quantifier strategy.

> Necessity Strategy: To prove a conclusion of the form of a necessity schema $\Box \phi$ we first set up a necessity sub proof whose conclusion is the operand ϕ within the scope of the necessity operator. If ϕ can be proven, $\Box \phi$ may be inferred by the rule of necessity introduction.

R18. <u>Necessity Elimination</u>: <u>From a modal schema consisting of a schema ϕ prefixed by a necessity operator we may infer ϕ itself</u>.

$$\frac{\Box \phi}{\phi}$$

When interpreted this rule permits us to infer from the necessity of a proposition to its truth.

<u>Rules Governing the Possibility Operator</u>. Again we have rules analogous to those for the quantifiers.

R19. <u>Possibility Introduction</u>: <u>From a schema ϕ we may infer</u> $\Diamond \phi$.

$$\frac{\phi}{\Diamond \phi}$$

This rule can be interpreted as permitting us to infer from the truth of a proposition to its possibility. If it is true that a man jumped seven feet high, it is surely also possible.

R21. <u>Possibility Elimination</u>: <u>From a modal schema of the form</u> $\Diamond \phi$ <u>and a necessity sub proof showing that ϕ leads to a modal schema ψ we may infer ψ</u>.

$$\Diamond \phi$$
$$\Box \begin{array}{|l} \phi \\ \vdots \\ \psi \end{array}$$
$$\overline{\psi}$$

where ψ must be a modal schema.

The requirement that ψ be a modal schema prevents an inference such as the following:

```
1 │ ◊A          P
2 │┌─A          P
3 │□│─A         2,r
4 │ A           1,2-3, ◊ elim (forbidden)
```

Because A expresses a possible proposition it does not follow that this proposition is true: it is logically possible but not true that Hubert Humphrey is the U.S. President in 1970.

An example follows of a proof employing the possibility rules.

```
1 │◊(A ⊃ B)           p
2 │┌A ⊃ B             p
3 │ │┌□A              p
4 │□│ │A              3, □ elim
5 │ │ │A ⊃ B          2,r
6 │ │ │B              4,5,m.p.
7 │ │ ◊B              6, ◊ intro
8 │ │□A ⊃ ◊B          3-7, ⊃ intro
9 │□A ⊃ ◊B            1,2-8, ◊ elim
```

To construct this proof we employ the

> Possibility Strategy: If a given step from which a modal schema ψ is to be inferred as conclusion is of the form ◊φ, then set up a necessity sub proof headed by the operand φ and attempt to prove ψ as its conclusion. If ψ can be proved, it can be inferred as the desired conclusion by the rule of ◊ elim.

This strategy is employed in constructing the necessity sub proof of steps 2-8 headed by A ⊃ B as its premiss. Since □A ⊃ ◊B can be proven within this sub proof, it may be inferred as the conclusion of the main proof by the rule of ◊ elim.

Conditional proofs in the modal calculus can be again regarded as proofs of derived rules to be used in later proofs. The conclusions of categorical proofs are, as before, expressions which when interpreted express logical truths in modal logic. An example follows of a categorical proof of a material equivalence.

```
 1  | □-A                   p
 2  |  | ◊A                 p
 3  |  |  | A               p
 4  |  |  | □ | -A           1,r
 5  |  |  |   | -□-A         3,4,-elim
 6  |  | -□-A               2,3-5, ◊ elim
 7  |  | □-A                1,r
 8  | -◊A                   2-7,-intro
 9  | -◊A                   p
10  |  | | A                p
11  |  | | ◊A               10, ◊ intro
12  |  | □ | -◊A            9,r
13  |  | -A                 10-12,-intro
14  | □-A                   10-13, □ intro
15  | □-A ≡ -◊A             1-8,9-14, ≡ intro
```

The first of the sub proofs needed to prove the conclusion is constructed by showing that ◊ A leads to a contradiction. In this proof we employ the Possibility Strategy to derive -□-A in step 6 from ◊ A. The second of the required sub proofs is constructed by proving □ -A in step 14 by means of the Necessity Strategy.

Iterative Modalities. It seems plausible to say that if the proposition expressed by a sentence A is necessary then it is necessary that A is necessary, or if □A, then □□A. Similarly, if A is possible, then it is necessary that it is possible, or if ◊ A then □◊A. To allow for the inferences □ A ⊢ □□A and ◊ A ⊢ □◊A we must modify, however, our reiteration principle. The inferences would be permissible if we adopt the following

Modified Principle of Modal Reiteration: Only schemata prefixed by a modal operator may be reiterated into a necessity sub proof.

The Modified Principle differs from our original Modal Principle in allowing the reiteration of possibility modal schemata and not requiring that the necessity operator be dropped after reiteration.

We shall call the modal calculus that results from adopting this Modified Principle the iterative modal calculus. It is evident that in this calculus the required conditional proofs to interative modalities can be accomplished.

```
1 | □A               p              1 | ◊A              p
2 | □ | □A           1,r            2 | □ | ◊A          1,r
3 | □□A              3, □ intro     3 | □◊A             2, □ intro
```

Step 2 in each proof is licensed by the Modified Principle, though they would be forbidden by our original Modal Principle.[2]

Whether the inferences $\Box A \vdash \Box\Box A$ and $\Diamond A \vdash \Box \Diamond A$ ought to be allowed within a modal calculus has been a subject of controversy. Certainly if a proposition such as that expressed by 'It is raining or not raining' is logically necessary, it is necessary that it is necessary. Similarly, if 'It is raining or cloudy' is possible, it is necessary that it is possible. Both seem to be a consequence of the fact that logical necessity and possibility are established by mechanical decision procedures. Yet the interpretations of the modal operators that occur together in such schemata as $\Box\Box A$ and $\Box\Diamond A$ seem quite different. If it is logically necessary that A, then it is necessary in some non-logical sense (perhaps the necessity of the synthetic a priori) that A is necessary. In general, it is mistaken to combine modal operators with different interpretations within the same inference. $\Box A \therefore \Box A$ is a valid inference form, but not if $\Box$ is interpreted in the premiss as the necessity expressed by a materially analytic sentence (e.g. 'Bachelors are unmarried') and as logical necessity in the conclusion. By restricting ourselves to our original Modal Principle of Reiteration we can avoid the dangers of giving modal operators different interpretations within the same inference.

Modal Operators and Quantifiers. So far we have restricted the application of the rules of the modal calculus to modal schemata whose operators have as their scopes sentence schemata. But the modal rules can also be applied to expressions with quantificational schemata occurring within the scopes of operators. We can provide categorical proofs within the modal calculus, for example, of both $\Box \forall x - (Px \land -Px)$ and $\Box \exists x(Px \land Qx) \supset \Diamond \exists xPx$ in the following manner.

```
1 | □ | x | | Px∧ -Px          p
2 |       | Px                 1,∧elim
3 |       | -Px                1,∧elim
4 |     | -(Px∧ -Px)           1-3,-intro
5 |   | ∀x-(Px∧ -Px)           1-4, ∀ intro
6 | □∀x-(Px∧ -Px)              1-5, □ intro
```

```
1 |   | □∃x(Px∧Qx)             p
2 |   | ∃x(Px∧Qx)              1, □ elim
3 |     | Px∧ Qx               p
4 | x |  | Px                  3, ∧ elim
5 |   |  | ∃ xPx               4, ∃ intro
6 |   |  | ◇ ∃xPx              5, ◇ intro
7 |   | ◇∃xPx                  2,3-6, ∃ elim
8 | □∃x(Px∧Qx) ⊃ ◇∃xPx         1-7, ⊃ intro
```

The proofs are constructed by means of modal rules plus rules for quantifiers as well as sentence connectives. Successive sub proofs are set up in accordance with a strategy for either a modal operator, a quantifier, or a sentence connective.

Indiscriminate combining of modal operators with quantifiers can lead to difficulties, however. Recall that in Section 31 we agreed to regard modal sentences in which 'necessary' and 'possible' occur as exclusively sentences of the meta-language in which object language sentences are mentioned. We excluded from later consideration the physical modalities applied to events or states of affairs. Modal schemata of the form $\Box \phi$ and $\Diamond \phi$ are thus to be taken as representing sentences of the meta-language. We can remain consistent with this agreement if we apply our calculus to modal schemata with quantificational schemata within the scope of the modal operator. A schema such as $\Box \forall x - (Px \wedge -Px)$ can be regarded as representing a modal sentence such as '"Nothing is both red and not red" is necessary', a meta-linguistic sentence in which the sentence 'Nothing is both red and not red' is mentioned.

But we cannot remain consistent with our agreement if we apply our calculus to quantificational schemata in which modal schemata occur within the scope of quantifiers. We might, for example, think that a proof of the following kind is possible.

1)
```
        1 |           | Px ∧ -Px      P
        2 |           | Px            1, ∧ elim
        3 | x    □    | -Px           1, ∧ elim
        4 |           -(Px ∧ -Px)     1-3, - intro
        5 |        □-(Px ∧ -Px)       1-4, □ intro
        6 | ∀x □-(Px ∧ -Px)           1-5, ∀ intro
```

In the proof the conclusion of the main proof $\forall x \Box -(Px \wedge -Px)$ follows from $\Box -(Px \wedge -Px)$ in step 5 if we regard this latter expression as an open schema in which x occurs free (the ϕv of the rule of $\forall$ intro).

But what kind of sentence could $\forall x \Box -(Px \wedge -Px)$ represent? Suppose that P represents the predicate '...is red'. Then we might take the quanitificational schema as representing the sentence,

    2) Everything is such that it is necessary that it is not both red and not red.

The open schema $\Box -(Px \wedge -Px)$ would then represent

    3) It is necessary that it is not both red and not red

with the pronoun 'it' being represented by the free variable. But if modal sentences are restricted to those of the meta-language as advocated in Section 31, then 3) is equivalent to

    4) 'It is not both red and not red' is necessary

in which the open sentence 'it is not both red and not red' is mentioned. Replacing 3) by 4), 2) becomes

    5) Everything is such that 'it is not both red and not red' is necessary.

But now 'it' is no longer a pronoun that can refer back to the subject term 'thing', and we have a sentence that no longer expresses a proposition. As it occurs in the middle of 5) the pronoun occurs as part of the quotation mark <u>name</u> of an open sentence. By mentioning the open sentence we have prevented the pronoun from performing its proper role. The result is a hybrid sentence the subject of which is an expression of the object language, while the remainder, what should be the predicate, is in the meta-language.[3]

No such mixing of levels of language occurs if $\forall x \Box -(Px \land -Px)$ is taken as representing the sentence,

 6) Everything is necessarily either red or not red

which predicates the property of being red or not red as necessary of everything (as an "essential property" in the Aristotelian terminology). Here necessity is a physical modality "<u>de re</u>" rather than an alethic modality "<u>de dicto</u>." As a sentence of the object language 6) contains no mentioned expression, nor is it equivalent to any other sentence containing one. As a result, no such problem as that confronting us for 2) as paraphrased into 5) arises. We need to prevent the formation of quantificational schemata of the form $\forall \mu \Box \phi \mu$ and $\exists \mu \Box \phi \mu$, therefore, only if these schemata represent meta-linguistic modal sentences.

To prevent their formation and thus restrict modal sentences to those of the modal language we can simply stipulate that no expression containing a modal operator can occur within the scope of a quantifier. To prevent such proofs as 1) we can also stipulate that the expressions $\phi \mu$ and $\phi \nu$ used in stating the quantifier rules may not be taken as representing schemata in which occur modal operators. In this way we prevent the derivation of the conclusion of 1) by means of the rule of $\forall$ intro.

<u>Substitution within Modal Sentences</u>. Difficulties also arise when we form schemata in which modal operators occur together with the identity sign. Consider the inference,

 7) The Morning Star is the Evening Star.
   It is necessary that the Morning Star is identical with the Morning Star.
   ∴ It is necessary that the Morning Star is identical with the Evening Star.

This is of the form

 8) $a = b$
   $\Box(a=a)$
   $\overline{\Box(a=b)}$

which is an instance of R16, the rule of identity elimination of Section 29. We derive the conclusion by substituting b for a in the second premiss. Yet 7) is clearly an invalid inference. The first premiss is true; the second is also, since any expression of the form a=a is analytically true. But the conclusion is false. It is not a necessary truth that the two singular terms 'the Morning Star' and 'the Evening Star' denote the same object, the planet Venus. This is instead a contingent proposition that had to be established by empirical investigation.[4]

To avoid justifying such invalid inferences as 7) with our identity rules a simple expedient is available. We can stipulate that in the rule of identity elimination, $\mu=\nu$, $\phi\mu \vdash \phi\nu$, $\phi\mu$ and $\phi\nu$ may not represent any schema in which a modal operator occurs. The effect of this restriction is to prevent the substitution within modal schemata of the sort found in the inference schema 8).[5]

Exercises.

I. Construct conditional proofs for the following.

1. $\Box(A \supset B) \vdash \Box A \supset \Box B$
2. $\Box(A \supset B), \Box -B \vdash \Box -A$
3. $\Box -A \vdash \Box(A \supset B)$
4. $\Box B \vdash \Box(A \supset B)$
5. $\Box A \lor \Box B \vdash \Box(A \lor B)$
6. $\Diamond(A \land B) \vdash \Diamond A \land \Diamond B$
7. $\Box(A \supset B), \Diamond -B \vdash \Diamond -A$
8. $\Box(B \supset -C), \Diamond(A \land B) \vdash \Diamond(A \land$
9. $\Diamond A \supset \Box B \vdash \Box(A \supset B)$
10. $\Box(A \equiv B) \vdash \Diamond A \equiv \Diamond B$

II. Construct categorical proofs for the following. Introduce and use derived rules based on previous proofs where convenient.

1. $\vdash \Box(A \supset A)$
2. $\vdash \Box\Box(A \supset A)$
3. $\vdash \Diamond \Box[(A \supset B) \supset (-B \supset -A)]$
4. $\vdash \Box(A \land B) \equiv \Box A \land \Box B$
5. $\vdash \Diamond(A \lor B) \equiv \Diamond A \lor \Diamond B$
6. $\vdash \Box A \equiv -\Diamond -A$
7. $\vdash \Diamond A \equiv -\Box -A$
8. $\vdash (\Diamond A \supset \Diamond B) \supset \Diamond(A \supset B)$
9. $\vdash -\Diamond(A \land B) \supset \Box(A \supset -B)$
10. $\vdash \Box(A \supset B) \equiv -\Diamond(A \land -B)$
11. $\vdash -\Diamond A \equiv \Box(A \supset -A)$
12. $\vdash \Box \forall x[-Px \supset (Px \supset Qx)]$
13. $\vdash \Box \forall x(Px \supset Qx) \land \Diamond \exists x Px \supset \Diamond \exists$
14. $\vdash \Box \forall x(Px \land Qx) \equiv \Box \forall x Px \land \Box \forall$
15. $\vdash \Box(a=b) \equiv \Box(b=a)$
16. $\vdash \Box(a=b) \land \Diamond(b=c) \supset \Diamond(a=c)$

NOTES FOR CHAPTER VII

Section 31.

1. The term 'alethic' is due to von Wright [Modal Logic], as are the classifications of the epistemic and deontic modalities that follow. In von Wright's classification, however, the physical modalities are a specific interpretation of the alethic. I have chosen to classify them as a separate branch because the objects to which the modal terms are applied are different.

2. The physical modalities can also be interpreted in terms of the "objective" probability of the occurrence of an event. A necessary event is one whose probability is 1, an impossible event one whose probability is 0, and a contingent event has a probability of some rational number between 1 and 0.

3. Included within this branch could also be included modal systems in which the modal trichotomies are interpreted in temporal terms. A necessary state of affairs can be regarded as one which is eternal, while a contingent state of affairs is one of finite duration, one with a beginning and end. The necessity of God's existence in one of Anselm's arguments seems to be this sense of necessity. For the development of temporal systems see Prior [Time].

4. For a clear statement of this view see Nagel [Ontology].

5. See Hintikka [Knowledge] for a system of epistemic modal logic and a discussion of its main issues.

Section 32.

1. Entailment (or strict implication) and compatibility were defined in this manner in Lewis and Langford [Logic], Ch. VI. Lewis' investigations of the properties of the strict implication relation is generally credited as marking the start of the modern development of modal logic. For a discussion of this contribution see the Kneales [Development], pp. 548ff. For a survey of the development of modal logic from Aristotle and the Medieval logicians through modern times see also the Kneales [Development] and Bochenski [History].

2. The distinction made here is between what Reichenbach calls the "relative" and "absolute" modalities. Cf. his [Logic], pp. 392ff.

3. The account of materially analytic sentences that follows is due to Carnap. Cf. his [Postulates].

4. The school of logical positivism maintained that analytic and synthetic a posteriori sentences constitute an exhaustive dichotomy of all meaningful sentences. For a clear statement of this view see Ayer [Language], Ch. IV. For a criticism of this view and a defense of the synthetic a priori see Pap [Semantics], Ch. 5. For a criticism of the analytic-synthetic distinction itself see Quine [Logical View], Essay II.

Section 33.

1. The decision procedure that follows is due to von Wright [Modal Logic].

2. For the construction of a calculus where the "paradoxes" are avoided by imposing relevance conditions on the premisses and conclusion of an inference see Anderson and Belnap [Entailment]. Lewis in [Logic] denied the paradoxical nature of 8) and 9). For support for Lewis' view see Bennett [Entailment].

Section 34.

1. The calculus that follows is due to Fitch [Logic], Ch. 3.

2. Alternate modal calculi were developed by Lewis and Langford [Logic]. What we are calling the iterative modal calculus is equivalent to Lewis' System S5. An axiomatic calculus in which the inference from $\Box A$ to $\Box\Box A$ is permissible but not from $\Diamond A$ to $\Box\Diamond A$ was called System S4. The modal calculus in which the necessity operator must be dropped is analogous to Lewis' System S2. For a survey of alternative calculi see Hughes and Cresswell [Modal Logic].

3. This is a restatement of Quine's criticisms of quantifying into modal contexts in [Notes] and [Logical View], Essay VIII. An expression such as $\Box$-(Px$\wedge$-Px) can be regarded as itself the meta-linguistic expression $\Box$'-(Px$\wedge$-Px)', rather than as <u>representing</u> a meta-linguistic sentence, as is done here. Prefixing the quantifier we get, Quine points out, $\forall x \Box$'-(Px$\wedge$-Px)' in which the quotation marks prevent the variables from lying within the scope of the quantifier.

4. Cf. Quine [Logical View], Essay VIII. The fact that for modal sentences such as the second premiss of 7) substitution of identicals does not preserve truth value Quine takes as evidence that modal sentences should be regarded as descriptive of the way we describe things, that is, sentences of the meta-language, rather than as descriptive of things and part of the object language: "necessity does not properly apply to the

fulfillment of conditions by objects (such as the ball of rock which is Venus ...) apart from special ways of specifying them." For objections to Quine's treatment of modalities see Marcus [Modalities].

5. Such a prohibition of substitution is not restricted to modal sentences. It is also necessary for belief sentences. The following inference, for example, is invalid:

>   Tom believes that Cicero denounced Cataline.
>   <u>Cicero is Tully.</u>
>   ∴ Tom believes that Tully denounced Cataline.

for that 'Cicero' and 'Tully' denote the same person may be unknown to Tom, and hence the premisses may be true and the conclusion false. Similar considerations hold where 'knows that...', 'hopes that...', 'wonders whether...', etc. replace 'believes that...'. Cf. Quine [Logical View], Essay VIII. Sentence contexts, both modal and non-modal, in which substitution fails to preserve the truth value of the original sentence are called <u>intensional contexts</u>, in contrast to <u>extensional contexts</u> to which the rule of identity elimination can be applied.

VIII. IMPERATIVES AND THE DEONTIC MODALITIES

35. Imperative Inferences

So far we have considered only indicative inferences, inferences whose constituent sentences are in the indicative mood and express propositions either true or false. Until quite recently logicians have restricted their attention exclusively to such inferences. But there are sentences in moods other than the indicative, among them sentences in the optative mood (e.g. 'Oh that I were young again!') and the imperative mood ('Open the window'). And it should not be surprising to find that there are inferences whose constituents are such sentences. Of special interest has proven to be imperative inferences, inferences whose premisses and conclusion are imperatives. These can be shown to have a logical structure similar to indicative inferences and to be susceptible to evaluation procedures similar to those of the sentence and predicate logic developed in earlier chapters.[1]

There seem to be two principal uses for imperative inferences. The first is in determining the consequences of a command or law that has been already issued or promulgated. One who has been ordered to perform a certain task may need to determine what is entailed by that task. For example, if he has been ordered to plant the seeds only if the ground is moist and the ground is, in fact, not moist, he should infer he is not to plant the seeds. A judge may also need to deduce the consequences of a law enacted by a legislature in order to apply the law to a specific case. The second principal use is made by those in the position of issuing commands or enacting laws. Whether or not a given command should be issued or a law enacted is often decided by deducing the consequences that would follow from obeying it and weighing their acceptability.

Sentence Radicals. In order to extend the procedures of sentence and predicate logic we must first distinguish between that aspect of a sentence that constitutes its logical subject-predicate structure and that which determines for us its mood. Consider the two sentences,

        1)   The letter is mailed
and 2)   Mail the letter.

1) is in the indicative mood, while 2) is in the imperative. The indicative sentence expresses a proposition that may be judged true or false; the imperative expresses what can be called a _prescription_ to be obeyed or disobeyed. It seems apparent that both have the same subject-predicate structure. The subject of

both is 'the letter'; in 1) it denotes the object upon which an action is to be performed. The predicates of 1) and 2) determine for us by the form of the verb and its position within the sentence, whether before or after the subject, the moods of the sentences. When we abstract from this mood determination we seem to be left with the verb 'to mail' as the common predicate of both. In 1) it expresses an attribute, in 2) an action. The common subject-predicate structure of 1) and 2) thus seems to be

    3)   The letter, to mail

a list of separate subject and predicate terms. The order of these terms and the form of the verb determines whether the sentence in which they occur as constituents is in the indicative or imperative mood. We shall call the subject-predicate structure of a sentence independent of mood determination the sentence-radical. 3) is thus the radical of both sentences 1) and 2). The order of constituent terms, form of verb, etc. by which mood is determined will be called the mood-determiner of a sentence.[2]

This analysis can be easily extended to sentences with more than one subject. The common sentence-radical of 'This letter is placed between the book and the desk' and 'Place the letter between the book and the desk' would be 'The letter, the book, the desk, to place between', with 'the letter', 'the book', and 'the desk' as subjects and the infinitive verb form 'to place between' as a triadic predicate. The position and form of the verb determines whether we have an indicative sentence expressing a proposition or an imperative expressing a prescription.

Of course, not all indicatives have corresponding imperatives from which we can abstract a common sentence-radical. There is no imperative, for example, corresponding to 'This book is red'. For there to be a corresponding imperative we must have a verb capable of expressing an action. Also, since an action prescribed by an imperative is to be performed in either the present or future, there can be no imperatives corresponding to indicatives whose verbs are in the past tense. Thus, there can be no imperative corresponding to 'The letter was mailed'. In contrast, we can construct a corresponding indicative reporting the performance of the prescribed action for every imperative simply by changing the verb from the active to passive voice and putting it in the present or future tense.

Now consider the pair of inferences,

    4)   The letter is mailed or it is burned.
            It is not mailed.
            ∴ The letter is burned.

> 5) Mail the letter or burn it.
> Don't mail it.
> ∴ Burn it.

4) is obviously a valid inference, since it is of the form of a disjunctive syllogism. 5) would also seem to be valid in the sense applicable to imperatives: it is impossible to obey the prescriptions expressed by its imperative premises without obeying that expressed by its conclusion. This validity of 5) seems due to the fact of its sharing a common logical form with 4) that is independent of the mood of the constituents of both inferences. It is definitions of the logical words 'or' and 'not' along with the recurrences of sentences in premises and conclusion in both inferences that determines this validity, not the mood of the recurring constituents. Replacing the sentences of both 4) and 5) by their common sentence radicals, we have the following combination of radicals:

> 6) The letter, to mail or the letter, to burn.
> The letter, to not mail.
> ∴ The letter, to burn.

This provides all the information needed to evaluate 4) and 5) as valid. Which mood determiners we add to the radicals in 6), whether indicative or imperative, will have no effect on validity.

Decision Procedures for Imperatives. The difference and similarity between inferences 4) and 5) can be represented by adopting a special logical symbolism. The general form of any sentence let us now represent by expressions of the form Am, where A represents the radical of the sentence and m its mood determiner. The determiner for the indicative mood we shall represent by the symbol '#', while the imperative mood-determiner will be represented by '!'. Both symbols are to be regarded as a type of logical constant distinct from the schema A standing for any arbitrary sentence-radical. The indicative 'The letter is mailed' will then be represented by the expression A#, while 'Mail the letter' will be represented by A!. The logical forms of 4) and 5) will be represented by

$$A\# \vee B\#$$
$$-A\#$$
$$\overline{B\#}$$
and
$$A! \vee B!$$
$$-A!$$
$$\overline{B!}$$

The form of 6) will be represented by simply dropping the mood-determiners in the inference schemata above, leaving

> 6) $A \vee B$
> $-A$
> $\overline{B}$

Our conclusion, then, is that it is by virtue of the form represented by 6) alone upon which the validity of both 4) and 5) depends.

This suggests that the same procedure employed in sentence logic for indicatives can be extended to imperative inferences. This is, in fact, the case. The prescription expressed by an imperative will be such that it will be either obeyed if the prescribed action is performed by the person to whom it is addressed or disobeyed if not performed. Obeying and disobeying a prescription can thus be regarded as values analogous to the truth and falsity of propositions. We shall refer to them as <u>performance values</u>. A valid imperative inference we have seen to be one for which it is impossible for someone to obey the prescriptions expressed by its premisses and disobey its conclusion. Validity in this sense can be determined by applying definitions that we give to the logical connectives. The negation of an imperative -A! will be disobeyed if A! is obeyed, and will be obeyed if A! is disobeyed. A disjunction of imperatives A! ∨ B! will be disobeyed if both disjuncts are disobeyed; otherwise, it is obeyed. A conjunction of imperatives such as 'Open the window and close the door' will be obeyed if both conjuncts are obeyed; otherwise it is disobeyed. The tables that follow (what we shall call <u>performance tables</u>) summarize these definitions.

A!	-A!
O	D
D	O

A!	B!	A! ∧ B!	A! ∨ B!
O	O	O	O
O	D	D	O
D	O	D	O
D	D	D	D

'O' should be taken here as the obeying of a prescription and 'D' for the disobeying of it. As we shall see in the next section, the definition of material implication (and thus material equivalence also) poses special problems.

Where we have the three connectives that can be defined in this manner, validity can easily be decided. The table for evaluating inference 5) would be

A!	B!	A! ∨ B!	-A!	B!
O	O	O	D	O
O	D	O	D	D
D	O	O	O	O
D	D	D	O	D

Since there is no interpretation under which B! is disobeyed when A! ∨ B! and -A! are obeyed, the inference is shown to be valid.

In like fashion we can decide which molecular imperative sentences express prescriptions that are impossible to disobey.

A prescription expressed by an imperative of the form A!∨ -A! (e.g. 'Close the door or don't close it') is clearly one of these, as can be shown again by a performance table.

<u>Extension to General Imperatives</u>. A decision procedure can also be constructed for imperative inferences whose validity depends on the recurrence of its constituent subject and predicate terms. We have seen how an imperative can be analyzed into a predicate expressing an action and one or more subjects denoting the objects relative to which the prescribed action is to be performed. Such analyzed sentences can be represented by the symbolism of predicate logic. Thus, 'Open the window' may be represented by Pa!, with P representing the predicate 'open...' expressing the prescribed action and a the subject 'the window' denoting the object upon which the action is to be performed. Similarly, 'Place the book on the table' may be represented by Rab!, with R representing the dyadic predicate 'Place...on...' and the individual schemata the two subjects. Since the imperative mood is determined by the form of the verb and its position within the sentence, the mood-determiner may also be written immediately after the predicate schema to produce P!a and R!ab. We shall adopt this form of representation because of its convenience in dealing with the problems raised in the next section.

Some general imperative sentences can be represented by this symbolism. Thus,

    7) Pick up everything
and 8) Pick up something

can be represented by ∀xP!x and ∃xP!x, with the variable x in the quantifiers representing the subject 'thing'. The representation of negative general imperatives poses special difficulties not found for indicatives. At first glance the negation of 7) would seem to be

    9) Don't pick up everything.

But 9) can be interpreted in two ways: either as a) the prescription that not everything be picked up, that is, that at least one thing be not picked up; or as b) the prescription that everything is not to be picked up. Very often we make interpretation b) explicit by changing the object of the verb to 'anything', while if a) is intended the stress in speaking 9) is on the quantifier 'every'. Which interpretation we give to 9) clearly determines its logical representation. If the first interpretation a) is given, the representation would be in terms of the "external" negation form -∀xP!x, the negation sign being prefixed to the quantifier. If interpretation b) is given, the representation is by the "internal" negation form ∀x-P!x. It is clear that only as interpreted in the first way can 9) be regarded as the negation of 7) in the sense of being its contradictory. Under interpretation b) 9) is only the contrary of 7). Similar considerations hold for the representation of 'Don't pick up something'. Interpreted as prescribing that nothing is to be picked up (where we would usually again say 'Don't pick up anything') its representation is $-\exists x P!x$, while interpreted as prescribing that something is not to be picked up we have $\exists x - P!x$. Under the first interpretation the sentence is the contradictory of 8), while the second interpretation gives us its contrary. The square of opposition diagrammed for indicative general sentences in Section 17 can thus be seen to hold also for general imperatives.[3]

The representation of relational imperatives is similar. For example, 'Place something on everything' would be represented by $\exists x \forall y R!xy$. The addition of the mood-determiner enables us to represent the imperative form.

Inferences between these general imperatives, e.g. one of the form $\forall x(P!x \wedge Q!x) \therefore \forall x P!x \wedge \forall x Q!x$ ('Pick up and burn everything. Therefore, pick up everything and burn everything'), can be evaluated as valid or invalid. The decision procedure for such inferences is analogous to that for predicate logic when applied to indicatives. When the domain denoted by the variables is finite, then universal and existential sentences are equivalent to conjunctions and disjunctions of singular imperatives, and the table method of sentence logic may be employed as is done in Section 20. When the domain is infinite (or so large as to make this method inconvenient) the method of distributive normal forms developed in Section 24 may be employed. Here the $\exists$-constituents for two predicates P and Q will be $\exists x(P!x \wedge Q!x)$, $\exists x(P!x \wedge -Q!x)$, $\exists x(-P!x \wedge Q!x)$, and $\exists x(-P!x \wedge -Q!x)$ whose values are those of the obedience or disobedience of actions to be performed relative to some member of the given domain.

Notice that we have so far restricted ourselves to general imperatives with category terms as subjects. Imperatives such as 'Pick up all the books' and 'Place at least one book on every table' have a more complex structure that will be discussed in the next section.

Addresses. There is a special difficulty in representing imperatives not present for indicatives. It arises from the fact that a prescription is obeyed or disobeyed by some person, and whether or not it is obeyed will depend on who performs the prescribed action. It will only be obeyed if the person performing this action is the one to whom a token of the imperative is addressed. For example, if an utterance of 'Close the door' is addressed to Peter, then to have Paul closing the door is not to have the prescription that is expressed obeyed. Only if Peter closes the door is it obeyed.

Let us call those elements of an imperative that indicate its addressee the <u>address</u> of the imperative. These can be left implicit and provided by context, as is often done in direct conversation. They can also be made explicit, as in 'Peter, close the door', where the proper name 'Peter' functions as the address, or in 'Someone pick up everything', with 'someone' as the address. Should addresses be included among the logical subjects of an imperative? It seems that they have a function different from that of a subject, and thus should not be. In 'Peter, close the door' 'Peter' does not denote the object upon which the action expressed is to be performed. And certainly we would not classify addresses as subjects when they occur in indicatives. The two indicatives 'Peter, the door is closed' and 'Paul, the door is closed' have different addresses, but the proposition expressed is regarded as the same for both. In the representation of the logical form of these sentences, then, addresses would not occur as subject terms.

Yet there are certain imperative inferences whose validity depends on logical relations between addresses. Here it does seem necessary to regard addresses as subjects, at least for the purposes of our logical representation. Consider, for example, the two inferences,

10) Peter, welcome Mary.
∴ Peter or Paul, welcome Mary.

11) Everyone, welcome Mary.
∴ Paul, welcome someone.

These are both valid, since it is impossible for their premisses to be obeyed without their conclusions being also obeyed. But in order to establish their validity we must represent the premisses and conclusions as relational sentences in which addresses are included as subjects. The logical forms of the inferences are thus respectively

$$\frac{R!ab}{R!ab \lor R!cb} \quad \text{and} \quad \frac{\forall x R!xb}{\exists y R!cy}$$

with R! for the imperative relational predicate '...welcome...', a and c for the addresses 'Peter' and 'Paul', b for 'Mary', and the variables x and y in the quantificational schema representing the general term 'person'. In these forms the inferences are now susceptible to our decision procedures.

Where the validity of inferences does not depend on logical relations between addresses, however, we should not represent them as subjects, assuming instead that the address remains constant for all constituent imperatives. In all but a few inferences this will be found to be the course taken.

Exercises. Evaluate the following inferences as valid or invalid. Notice that in inferences 1-6 the recurring elements are sentences, while in 7-10 they are terms. Take care to represent addresses only where essential to the evaluation of the inference.

1. Don't let the visitor in and not take his hat. Let the visitor in. Therefore, take his hat.
2. Don't let the visitor in and don't take his hat. Let the visitor in. Therefore, take his hat.
3. Don't either let the visitor in or take his hat. Therefore, don't let the visitor in.
4. Alfred, either don't let the visitor in or don't take his hat. Therefore, don't take the visitor's hat.
5. Don't let the visitor in, Alfred, and don't ask his name or address. But don't fail to let the visitor in. Therefore, make him feel at home.
6. Either bring back a reply to the message immediately or wait there and return later. But don't return later. Therefore, don't wait there either.
7. Don't either burn or destroy anything. Therefore, either don't burn anything or don't destroy anything.
8. Either deliver something, Alfred, or bring something back. Don't deliver everything. Therefore, don't bring something back.
9. Peter and Paul, don't injure anyone. Therefore, Paul, don't you injure Bill.
10. Peter or Paul, invite everyone inside. Peter, don't you invite Bill inside. Therefore, Paul, you invite someone inside.

36. Mixed Inferences

In the previous section we considered those inferences whose constituents are all in the imperative mood. By considering inferences whose constituents are in both the imperative and indicative moods, or mixed inferences, we greatly enlarge the scope of our evaluation procedure. An example of a mixed inference would be

> 1) If it is raining, close the windows.
> It is raining.
> ∴ Close the windows.

Here the antecedent of the conditional that constitutes the first premiss is in the indicative mood, while its consequent is an imperative. This inference we would assess as valid in the sense applicable to mixed inferences: it is impossible to obey the conditional imperative and for the second indicative premiss to be true, without obeying the conclusion. But to establish a decision procedure for such inferences poses special difficulties.

Conditional Imperatives. Most of these difficulties stem from the nature of the conditional imperatives that are constituents of most mixed inferences. Such conditionals are a common form of imperative. In issuing a command to someone we often specify the circumstance in which the prescribed action is to be performed. We may command someone to slow his car if the streets are slippery, to use a lever if a load is too heavy, etc. The sentences used to issue such commands are conditional imperatives whose indicative antecedent expresses a proposition specifying the circumstance under which the action expressed by the consequent is to be performed. Let us use the sentence schema A without a mood-determiner to now represent the indicative antecedent, omitting # for the sake of convenience. Then if B! represents the imperative consequent, the form of conditional imperatives generally is A⊃B!.

It is important to notice that this is also the only form conditional imperatives can take, since no conditional can have an imperative antecedent. We can say 'If you close the door, open the window' in which the antecedent reports the performance of an action, but not 'If close the door, open the window'. The role of the antecedent of a conditional imperative is that of stating a possible circumstance. It is obvious that this can only be performed by an indicative and never by an imperative prescribing an action. As a consequence every inference in which a conditional imperative occurs is required to be a mixed inference.[1]

What kind of truth table definition can be given for the implication sign that occurs in the expression A⊃B!? The reasoning followed for indicatives in Section 7 for indicative implication seems to dictate the following definition:

A	B!	A⊃B!
T	O	O
T	D	D
F	O	O
F	D	O

in which are listed four possible combinations of the truth or falsity of the indicative antecedent and the obedience or disobedience of the imperative consequent. As before for the material implication sign defined for indicatives, the first two interpretations of the table pose no problem. If the antecedent is true, then if the consequent is obeyed the conditional is obeyed, while if the consequent is disobeyed so is the conditional. The last two interpretations seem only capable of being justified in conjunction with the principle of bivalence for imperatives. If the state of affairs described by the antecedent fails to obtain and the antecedent is false, then surely the prescription expressed by A⊃B! cannot be disobeyed. But if a prescription must be either obeyed or disobeyed, it follows the conditional is obeyed.

This definition of the material implication sign for conditional imperatives requires us to abandon certain definitions available for indicatives. First of all, it does not seem possible to define the material implication sign used to represent conditional imperatives in terms of negation and disjunction or negation and conjunction. The definition $A \supset B! =_{df} -A \vee B!$ is not available, since there seems no way of defining the disjunction sign when one disjunct is an indicative and the other an imperative. (What value does $A \vee B!$ take when A is true and B! is obeyed?) Nor is $A \supset B! =_{df} -(A \wedge -B!)$ available for the same reason.

It is impossible also to define material equivalence by material implication with $A \equiv B! =_{df} (A \supset B!) \wedge (B! \supset A)$, since the second conjunct of the definiens contains an imperative antecedent. The definition $A \equiv B! =_{df} (A \supset B!) \wedge (-A \supset -B!)$ does seem available, however. To tell someone to do B if and only if A is to tell him to do B if A and to not do B if A is not the case. The truth table definition of $A \equiv B!$ is thus

A	B!	A≡B!
T	O	O
T	D	D
F	O	D
F	D	O

There is still another departure from the form of indicatives that must be noted. There seems to be no imperative corresponding to the form $-(A \supset B!)$ in natural languages. It might be taken to represent a sentence such as 'Don't if it is raining open the windows'. But this is synonymous with 'If it is raining don't open the windows' and hence should be represented by $A \supset -B!$. Schemata of the form $-(\phi \supset \psi)$, where ϕ is an indicative and ψ an imperative should thus be excluded from our symbolic language. Also to be excluded is a schema such as $-(A \equiv B!)$, since 'If and only if it is raining don't open the windows' would be represented by $A \equiv -B!$.

With performance table definitions of material implication and equivalence it is possible to evaluate mixed inferences in a table with both truth and performance values. Inference 1), for example, is evaluated by the following table:

A	B!	A⊃B!	A	B!
T	O	O	T	O
T	D	D	T	D
F	O	O	F	O
F	D	O	F	D

Since it is impossible for the conditional imperative to be obeyed, its indicative antecedent to be true, and the imperative B! to be disobeyed, the mixed inference is shown to be valid. Notice that mixed inferences such as 1) with an imperative as a premiss cannot be put in the form of a single conditional sentence as in indicative logic (cf. Section 8). The conditional (A⊃B!)∧A⊃B! violates the restrictions that there can be no mixed conjunctions and no antecedents of imperative conditionals in the imperative mood.

General Imperatives within Mixed Inferences. There are also mixed inferences in which general sentences occur either as premisses or conclusion, e.g.

> 2) Pick up every box.
> This is a box.
> ∴ Pick this up.

To evaluate such an inference we need a way of representing general imperatives that will disclose the logical connection between premisses and conclusion.

This representation is made possible by regarding a general imperative such as

> 3) Pick up every box

as expressing a prescription on the condition of a certain proposition being true. We can paraphrase 3) as

> 4) Everything is such that if it is a box, then pick it up.

Introducing the universal quantifier with its variable representing the subject 'thing' and the material implication sign we have

> 5) $\forall x(x \text{ is a box} \supset \text{pick } x \text{ up})$.

Finally we can obtain the quanitificational schema

> 6) $\forall x(Px \supset Q!x)$

in which P represents the predicate in the indicative mood '...is a box' and Q! the imperative predicate 'pick...up' expressing the prescribed action.

An existential imperative such as

> 7) Pick up some box

may be paraphrased as

> 8) There is at least thing such that it is a box and pick it up.

Its representation is now

> 9) $\exists x(Px \wedge Q!x)$

Note that in 9) the role of the first conjunct, like that of the antecedent in 6), is to specify the kind of object upon which the prescribed action is to be performed. The conjunction sign can thus be used in an open schema within the scope of an existential quantifier, although we have seen that there are no grounds for assigning a mixed conjunction $A \wedge B!$ either a truth or performance value.

The problem of negative imperatives noted in the previous section arises also for complex general imperatives. The negative sentence,

> 10) Don't pick up every box

can be either interpreted as prescribing that not every box is to be picked up and represented by $-\forall x(Px \supset Q!x)$ or as prescribing that every box is not to be picked up and represented by $\forall x(Px \supset -Q!x)$. In the latter case it is more usual to say 'Don't pick up any box'. Similarly,

> 11) Don't pick up some box

can be represented by the external negation form $-\exists x(Px \wedge Q!x)$ or by the internal negation form $\exists x(Px \wedge -Q!x)$, depending on whether no box is to be picked up or some box is not to be picked up. In the former case it is more usual again to say 'Don't pick up any box'. Logical equivalences hold between these quantificational schemata: $\exists x(Px \wedge -Q!x)$ is logically equivalent to $-\forall x(Px \supset Q!x)$, and $\forall x(Px \supset -Q!x)$ to $-\exists x(Px \wedge Q!x)$.

With such quantificational schemata available the representation of 2) becomes

$$\frac{\forall x(Px \supset Q!x)}{Q!a}$$

It is obviously valid, since from $\forall x(Px \supset Q!x)$ we may infer $Pa \supset Q!a$. An inference whose premisses and conclusion are all general sentences, such as one of the form

$$\frac{\forall x(Px \supset R!x)}{\forall x(Px \wedge Qx \supset R!x)}$$

could be evaluated by the distributive normal form method of monadic predicate logic outlined in Section 23. The distributive normal form of its premiss is $-[\exists(PQ\bar{R}!) \vee \exists(P\bar{Q}\bar{R}!)]$, while that of its conclusion is $-\exists(P\bar{Q}\bar{R}!)$. The appropriate truth table would show that it is impossible to obey the former and disobey the latter, and hence that its inference form is valid.

Mood Constancy. There are several restrictions on the types of imperatives that can be formed from constituents of different moods. As we saw, neither a conjunction whose conjuncts are of different moods nor a disjunction with differing disjuncts are permissible molecular sentences, since we cannot state conditions under which they take values relative to the values of their constituents. Also, we saw that a conditional imperative must have an indicative antecedent, and hence that a biconditional can have at most one imperative constituent and can only be defined in terms of conditionals with indicative antecedents.

There is also a restriction on the manner in which constituents of different moods occur within a mixed inference. The restriction is that the constituent elements upon which the validity of the inference depends, whether sentences or terms, must be in the same mood at every occurrence within the inference. We can state this as the following

> Principle of Mood Constancy: The constituent elements of an inference must be in the same mood for every occurrence.[2]

We shall call inferences valid by virtue of a connection between their constituent radicals alone inferences having radical validity. We saw in the previous section that all valid "pure" indicative and imperative inferences have radical validity, and that this is alone required for their validity. To be valid a mixed inference must also have radical validity. But it must in addition satisfy the Principle of Mood Constancy. The following is an example of an inference having radical validity, but nevertheless invalid.

> 12) If the door is closed, open the window.
> Don't open the window.
> ∴ Don't close the door.

Here the radical 'to close, the door' recurs as the antecedent of the conditional and as the conclusion, but within a sentence in the indicative mood in the former case and within an imperative mood-determiner in the latter. 12) is clearly an invalid inference. It is not necessary that if the first two premisses are obeyed that the conclusion is obeyed also, since the first premiss might be obeyed if the door were accidentally closed (rather than being closed by the person to whom the imperative conclusion is addressed). This invalidity is due to a violation of the Principle of Mood Constancy, since as an instance of modus tollens 12) has radical validity. By changing the conclusion from the imperative to the indicative mood and satisfying the Principle we would change the inference to a valid one. If the premisses of 12) were both obeyed it must follow that the door is not closed, though not that any one had left it open.

The same requirement also holds for the recurring predicates evaluated within predicate logic. The inference,

> Pick up everything that has been opened.
> This box is not picked up.
> ∴ This box has not been opened.

would seem to be of the form

13) $\forall x(Px \supset Q!x)$
$\underline{-Qa}$
$-Pa$

where P represents '...is opened' and Q 'pick up...'. From 13) it is apparent that the predicate 'pick up...' recurs in different moods in the first and second premisses. Again, the inference is invalid because of this shift of mood. If we change either the first occurrence of Q to the indicative mood or the second occurrence to the imperative mood ('Don't pick up this box'), we satisfy the Principle of Mood Constancy and change an invalid inference to a valid one.

Despite the contrast between the logical structures of imperatives and indicatives, it would be misleading to speak of a branch of logic called "imperative logic" distinguished from "indicative logic". Instead, it seems more accurate to speak of a common sentence and predicate logic applicable to sentence-radicals independently of mood. When this logic is applied to sentences in the indicative mood we have as results the decision procedures developed in Chapters II and V. It can also be applied in the manner outlined in this and the preceding section to imperatives. Its application here, however, is much narrower due to the special nature of imperative conditionals and general imperatives and to the further restrictions on validity imposed by the requirements of the Principle of Mood Constancy.

Exercises. Evaluate the following inferences. If an inference is invalid, indicate the reason for its invalidity, whether radical invalidity or failure to satisfy the Principle of Mood Constancy. Note that the recurring elements of inferences 7-10 are predicates.
1. If the ground is moist, Elmer, plant the seeds. Don't plant the seeds. Therefore, the ground is not moist.

2. Don't plant the seeds unless the ground is moist and there is warm weather. The ground is moist. Therefore, plant the seeds.
3. Plant the seeds only if you prepare the ground with fertilizer. Don't prepare the ground with fertilizer. Therefore, don't plant the seeds.
4. If you earn over $700 a year you must file an income tax return if a U. S. citizen. You are a U. S. citizen. Therefore, if you don't earn over $700 a year don't file an income tax return.
5. You must file a return unless you earn less than $700. If you file a return, use Form 1040. Therefore, if you earn less than $700, use Form 1040.
6. Construction will either be completed or be still in progress. If it is still in progress don't cross the bridge. But if it is completed, don't cross the bridge either. Therefore, don't cross the bridge.
7. Deliver all the packages. Therefore, don't deliver anything unless it is a package.
8. Don't pick up any strangers. Some children are strangers. Therefore, don't pick up some children.
9. No motor vehicles are allowed in the park. All motor scooters are motor vehicles. Therefore, no scooters are allowed in the park.
10. Don't drive any motor vehicles in the park. Some bicycles are motor vehicles. Therefore, some bicycles are not to be driven in the park.
11. Don't read any novel on the shelf. All novels on the shelf are books. There is a novel on the shelf. Therefore, don't read some book.
12. Don't read every book. Any book not read is to be sold. Therefore, sell at least one book.

37. The Calculus Applied to Imperatives

To apply the natural deduction calculus developed in Chapters III and VI to imperatives requires imposing some restrictions on the types of schemata to which the rules can be applied and minor modifications of two rules. The restrictions and modifications arise from the special features of imperatives just discussed. As before, conditional proofs by means of the calculus provide an alternative means of establishing the validity of imperative inferences.

The Sentence Calculus. The only modifications in the rules governing the sentence connectives are those required by the special nature of conditional imperatives. The rules governing the negation, conjunction, and disjunction connectives remain the same, with the meta-schemata ϕ, ψ, and χ now taken as representing arbitrary imperative schemata. For example, the proof of the rule of disjunctive syllogism would be constructed as follows:

```
 1  A! ∨ B!              p
 2  -A!                  p
 3    │ A!               p
 4    │   │ -B!          p
 5    │   │ A!           3,r
 6    │   │ -A!          2,r
 7    │ --B!             4-6,-intro
 8    │ B!               7,--elim
 9    │ B!               p
10    │ B!               9,r
11  B!                   1,3-8,9-10, ∨ elim
```

The only difference between this proof and the corresponding proof applied to indicative schemata is that the rules of disjunction elimination, negation introduction, and double negation elimination are applied to imperative schemata.

In applying the rules in this manner, however, we must be careful to adhere to the Principle of Mood Constancy and not let a given meta-schema represent both an indicative and imperative schema in an application of the rule. For example, we cannot infer A from A!∧B! by means of the rule of conjunction elimination, $\phi \wedge \psi \vdash \phi$, since ϕ would be applied to an imperative in the premiss and an indicative in the conclusion. Also, no step in a proof may consist of a disjunction or conjunction whose constituents are in different moods. This restriction prevents us from applying the negation, conjunction, and disjunction rules to such schemata as A∧B!, A!∨B, or -(A∧B!), which, as we have seen, are incapable of either a truth or performance table interpretation.[1] We cannot, therefore, infer A∨B! from A by disjunction introduction nor A∧B! from A and B! by conjunction introduction.

In contrast, when implication rules are applied to imperative schemata we must allow application of the rules to schemata of different moods, with suitable restrictions being imposed in order that our rules remain valid. We saw in the previous section that an imperative schema must have an indicative antecedent and imperative consequent because of the nature of the imperative conditionals they represent. As a result a meta-schema of the form $\phi \supset \psi$ occurring in a rule applied to imperative schemata must be understood in such a way that ϕ represents an indicative schema and ψ an imperative. This together with the Principle of Mood Constancy requires R1, the rule of implication introduction, to be understood as stating that if we can derive an imperative conclusion ψ from an indicative premiss ϕ we can infer an imperative of the form $\phi \supset \psi$. Similarly, R2, the rule of <u>modus ponens</u>, must be understood as stating that from an indicative schema ϕ and an imperative schema $\phi \supset \psi$ we can infer the imperative schema ψ.

The application of the implication rules applied in this manner may be illustrated by the following proof:

```
1 | A∧B ⊃ C!              p
2 |   | A                  p
3 |   |   | B              p
4 |   |   | A              2,r
5 |   |   | A∧B            4,3, ∧ intro
6 |   |   | A∧B ⊃ C!       1,r
7 |   |   | C!             5,6, m.p.
8 |   | B ⊃ C!             3-7, ⊃ intro
9 | A ⊃ (B ⊃ C!)           2-8, ⊃ intro
```

The inference to the imperative C! in step 7 is permissible, since A∧B is an indicative, while step 6, A∧B⊃C!, is an imperative implication with an indicative antecedent. It is permissible to derive the imperative implication in step 8, since the premiss of the sub proof from which it is derived by implication introduction is an indicative and its conclusion an imperative. A similar justification can be given for the conclusion of the sub proof.

The restrictions just specified on the form of schemata to which the rules can be applied have the effect of disallowing proofs that are possible for indicatives. It is impossible, for example, to derive the rule of contraposition for imperatives in the following manner:

```
1 | A ⊃ B!                 p
2 |   | -B!                 p
3 |   |   | A               p
4 |   |   | A ⊃ B!          1,r
5 |   |   | B!              3,4, m.p.
6 |   |   | -B!             2,r
7 |   | -A                  3-6, -intro
8 | -B! ⊃ -A                2-7, ⊃ intro (forbidden)
```

Because the premiss of the first sub proof is an imperative and its conclusion an indicative it is impossible to derive the conclusion in step 7. Another effect of the restrictions imposed on imperative schemata is to rule out use of the rule of implication introduction to construct categorical proofs of imperative implications. These are ruled out because it is impossible to derive an imperative conclusion from an indicative antecedent alone.

Recall from the previous section that A≡B! was defined by (A⊃B!) ∧ (-A⊃-B!) in order to prevent an imperative antecedent in an implication schema. As a consequence of this definition we must revise the two rules governing the material equivalence connective when they are applied to imperatives. The two rules now become

R9'. **Equivalence Introduction for Imperatives**: From two sub proofs, one leading from an indicative schema ϕ to an imperative conclusion ψ, the other from an indicative -ϕ to an imperative -ψ, we may infer a material equivalence of the form ϕ≡ψ applicable to imperatives.

```
1 | ϕ
  | ·
  | ·
  | ·
  | ψ
  |─────
  | -ϕ
  | ·
  | ·
  | ·
  | -ψ
  ──────
   ϕ ≡ ψ
```
where ϕ represents an indicative schema and ψ an imperative schema.

R10'. **Equivalence Elimination for Imperatives**: From a material equivalence ϕ≡ψ where ϕ represents an indicative schema and ψ an imperative schema we may infer ϕ⊃ψ and -ϕ⊃-ψ.

$$\frac{\phi \equiv \psi}{\phi \supset \psi}$$
$$-\phi \supset -\psi$$

where ϕ represents an indicative schema and ψ an imperative.

From the restrictions written into R9' it is evident that it cannot be used to construct categorical proofs. We are thus left with only the negation rules as means of constructing such proofs when we apply the sentence calculus to imperatives.

The Predicate Calculus. The quantifier rules can be applied to imperative quantificational schemata without modification. The following proof illustrates the application of the rules governing the universal quantifier.

```
1 | ∀x(Px ⊃ R!x)           p
2 |   | Px ∧ Qx             p
3 |   | Px                  2, ∧ elim
4 | x | Px ⊃ R!x            1, r, ∀ elim
5 |   | R!x                 3, 4, m.p.
6 | Px ∧ Qx ⊃ R!x           2-5, ⊃ intro
7 | ∀x(Px ∧ Qx ⊃ R!x)       6, ∀ intro
```

The conclusion is derived by applying the rule of universal quantifier introduction on the imperative open schema in step 6.

The application of the existential quantifier rules is similar, as illustrated by the proof of ∃xQ!x from ∃x(Px ∧ Q!x).

```
1  | ∃x(Px∧Q!x)          p
2  |⎡ Px∧Q!x             p
3  |⎢x ⎡Q!x              2, ∧ elim
4  |⎢  ⎣∃xQ!x            3, ∃ intro
5  |∃xQ!x                1,2-4, ∃ elim
```

Note that the inference from step 2 to 3 involves an application of the rule of conjunction elimination to schemata of different moods. This application of the rule can be allowed within general sub proofs used with the rule of ∃ elim, since open schemata within such proofs are not interpreted as either true or false or obeyed or disobeyed.

With the quantifier rules we can easily establish the validity of inferences whose premisses and conclusion include general imperatives. For example, the imperative "syllogism,"

> Tie down everything that can be washed overboard.
> Some deck chairs can be washed overboard.
> ∴ Tie down some deck chairs.

is of the form

$$\forall x(Qx \supset R!x)$$
$$\underline{\exists x(Px \land Qx)}$$
$$\exists x(Px \land R!x)$$

with Q for '...can be washed overboard', R! for 'tie down...', and P for '...is a deck chair'. It is shown to be valid by the following conditional proof:

```
1  | ∀x(Qx⊃R!x)          p
2  | ∃x(Px∧Qx)           p
3  |⎡ Px∧Qx              p
4  |⎢ Qx                 3, ∧ elim
5  |⎢ Qx⊃R!x             1,r, ∀ elim
6  |⎢ R!x                4,5, mp
7  |⎢ Px                 3, ∧ elim
8  |⎢ Px∧R!x             7,6, ∧ intro
9  |⎣ ∃x(Px∧R!x)         8, ∃ intro
10 | ∃x(Px∧R!x)          2,3-9, ∃ elim
```

The application to relational imperatives poses no special problems. For example, we can show the validity of the inference,

> Contribute some money only to Democratic candidates.
> Spiro Agnew is not a Democratic candidate.
> ∴ Don't contribute some money to Spiro Agnew.

of the form

$$\exists x[Px \land \forall y(-Qy \supset -R!xy)]$$
$$\underline{-Qa}$$
$$\exists x(Px \land -R!xa)$$

with P for '...is money', Q for '...is a Democratic candidate', R! for 'contribute...to...', and a for 'Spiro Agnew'. Its proof is

```
1  │ ∃x[Px ∧ ∀y(-Qy ⊃ -R!xy)]         p
2  │ -Qa                               p
3  │ ┌ Px ∧ ∀y(-Qy ⊃ -R!xy)            p
4  │ │ ∀y(-Qy ⊃ -R!xy)                 3, ∧ elim
5  │ │ -Qa ⊃ -R!xa                     4, ∀ elim
6  │ │ -Qa                             2, r
7  │x│ -R!xa                           5,6, m.p.
8  │ │ Px                              3, ∧ elim
9  │ │ Px ∧ -R!xa                      8,7, ∧ intro
10 │ └ ∃x(Px ∧ -R!xa)                  9, ∃ intro
11 │ ∃x(Px ∧ -R!xa)                    1,3-10, ∃ elim
```

The rule of universal quantifier elimination is applied to step 4 with the substitution of an individual schema for the free variable.

We can also apply the identity rules to imperative schemata. This application may be illustrated by the inference,

> Describe every American President.
> Thomas Jefferson was an American President.
> Thomas Jefferson was the third American President.
> ∴ Describe the third American President.

whose form is

$$\forall x(Px \supset Q!x)$$
$$Pa$$
$$\underline{a=b}$$
$$Q!b$$

with P for '...is an American President', Q! for 'describe...', a for 'Thomas Jefferson', and b for 'the third American President'. The proof of its validity is

```
1 │ ∀x(Px ⊃ Q!x)      p
2 │ Pa                p
3 │ a=b               p
4 │ Pb                2,3, =elim
5 │ Pb ⊃ Q!b          1, ∀ elim
6 │ Q!b               4,5, m.p.
```

Exercises.

I. Construct conditional proofs for the following by means of the modified rules of the predicate calculus. Use derived rules based on previous proofs where convenient.

1. $B \supset C! \vdash A \supset (B \supset C!)$
2. $A \supset (B \supset C!) \vdash (A \supset B) \supset (A \supset C!)$
3. $A, -A \vdash B!$
4. $-A \vdash A \supset B!$
5. $A \supset B!, -B! \vdash -A$
6. $A \supset B! \vdash A \wedge C \supset B!$
7. $A \vee B, A \supset C!, B \supset D! \vdash C! \vee D!$
8. $-(A! \vee B!) \vdash -A! \wedge -B!$
9. $-(A! \wedge B!) \vdash -A! \vee -B!$
10. $A \equiv B! \vdash -A \equiv -B!$
11. $\forall x P!x \vee \forall x Q!x \vdash \forall x (P!x \vee Q!x)$
12. $\forall x (Px \supset Q!x) \vdash \forall x Px \supset \forall x Q!x$
13. $\exists x (P!x \wedge Q!x) \vdash \exists x P!x \wedge \exists x Q!x$
14. $\forall x (Px \supset Q!x), \exists x Px \vdash \exists x (Px \wedge Q!x)$
15. $\forall x (A \supset P!x) \vdash A \supset \forall x P!x$
16. $\exists x (A \wedge P!x) \vdash A \wedge \exists x P!x$
17. $\forall y R!ay \vdash \exists x \exists y R!xy$
18. $\forall x (Px \supset \forall y R!xy) \vdash \forall x Px \supset \exists x \exists y R!xy$

II. Demonstrate the validity of the following inferences by means of the predicate calculus. Note that for 7-10 this requires representing addresses such as 'everyone', 'someone', and 'Tom' occurring in the inferences.

1. Inferences 2,3,5,9, and 10 of the exercises at the end of Section 35.
2. Inferences 1,6, and 9-11 of the exercises at the end of Section 36.
3. Pick up every piece of furniture. This desk is a piece of furniture. Therefore, pick up this desk.
4. Don't pick up any piece of furniture. This desk is a piece of furniture. Therefore, don't pick it up.
5. Go to the largest store in town. The super market on Wall Street is the largest store in town. Therefore, go to the super market on Wall Street.
6. Go from the largest store in town to the railroad station. The largest store in town is its only super market. The railroad station is the building on the corner of Wall and 13th. Therefore, go from the town's only super market to the building on the corner of Wall and 13th.
7. Tom, you pick up some trash, since everyone is to pick up some trash.
8. Every worker in this plant must report to Smith. Jones is a worker in the plant. Therefore, Jones, you must report to someone.
9. Every platoon leader is to study the operations map. Some lieutenants are platoon leaders. Therefore, some lieutenants are to study something.

10. No motorist is to ignore any warning. All stop signs are warnings. Therefore, no motorist is to ignore any stop sign.

38. Deontic Sentences

As we noted in Section 33, an action may be characterized as being either obligatory, forbidden, or indifferent, that is, neither obligatory nor forbidden. We turn now to deontic logic where inferences are evaluated whose validity depends on formal relationships between the terms of this modal trichotomy. Before considering the decision procedure used here, however, we must first discuss the relationships between imperative sentences used to prescribe actions and deontic sentences used to assess the modal status of these actions.

Imperative and Deontic Sentences.

We may say to another

1) Pay the bill.

We may also say to him

      2) You ought to pay the bill

or   3) It is obligatory that the bill be paid (by you).

1) is a sentence in the imperative mood expressing a prescription. 2) and 3) are deontic sentences in the indicative mood expressing a proposition that is either true or false. If the person addressed is indeed under the obligation the proposition is true; otherwise, it is false. Although they differ in mood, however, both the imperative and the deontic sentences require an action of a person. What then is the difference between saying 1) to someone and saying 2) or 3)? The answer seems to be that the latter imply that some reason or justification can be given for requiring of the person the action of paying the bill. It is implied either that the person issuing the command is in a position of authority over him, or that the law requires his obedience, or some similar reason. In contrast, no justification for the requirement is implied by the imperative; it may be used to issue a command which rests only on the arbitrary whim of the person who issues it. This difference may be put as follows: the deontic sentence 2) 'You ought to pay the bill' rightly evokes the question 'Why?', and the question deserves an answer; the imperative 'Pay the bill' often does not evoke a why-question and if the question is forthcoming may be inappropriate and deserve only the reply 'Because I told you to'. Often it makes no sense to speak of a reason for obeying an imperative other than the immediate fear of the sanction that may follow disobedience. If we do what someone tells us we ought to do, on the other hand, it is invariably on the basis of some reason other than immediate fear.[1]

Recall from Section 33 that a modal sentence such as 'It is necessary that bachelors are unmarried' is to be regarded as equivalent to the meta-linguistic sentence '"Bachelors are unmarried" is necessary' in which a sentence is mentioned. It seems that deontic sentences such as 3) should also be regarded as disguised sentences of the meta-language. When we say to someone 'It is obligatory that you pay the bill', we do not seem to be using the deontic sentence to describe a state of affairs. Questions of what people ought to do, as moralists have emphasized, are not questions about what is the case. We are instead characterizing with a deontic sentence an action to be performed. Such an action would be performed when the prescription expressed by an imperative is obeyed. For example, to obey the imperative 'Pay the bill' is to perform the action of paying the bill. The deontic sentence 3) can therefore be regarded as equivalent to the meta-linguistic sentence,

    4) 'Pay the bill' is obligatory

in which the imperative 1) is mentioned. As before for the alethic modalities, it is not the imperative sentence that is being characterized as obligatory in 4). Rather, it is the action of paying the bill which constitutes obeying the prescription expressed by the imperative.[2]

Similar considerations hold for deontic sentences of other forms. The deontic sentences,

    5) It is forbidden to speed on a highway
and 6) It is permissible to walk on the grass

should be regarded as equivalent to

    7) 'Speed on a highway' is forbidden
and 8) 'Walk on the grass' is permissible.

What is being characterized in 7) and 8) as forbidden and permissible are again the actions of speeding and walking that would be performed if the prescriptions expressed by the quoted imperative sentences were obeyed, and not the sentences themselves. 7) and 8), like 4), are admittedly not sentences that we encounter in ordinary language. They are artificial sentences that nevertheless provide us with meta-linguistic versions of 5) and 6).

The distinction between imperatives expressing prescriptions and deontic sentences expressing propositions is somewhat complicated by the use of the words 'must' and 'may'. As we have seen in the exercises of the previous sections of this chapter, 'must' can be used in order to issue a command. To say 'You must pick up this book' is usually to command someone to perform the action of picking up the book. But 'must' can also be used in the sense of 'ought to' and occur within a deontic sentence. To say 'You must pay the bill' can thus be taken in the sense of 'You ought to pay the bill' or 'You are (legally) obligated to pay it'. Here we would have a deontic sentence expressing a proposition. Similarly, 'may' can be used to grant permission to someone, as in 'You may open the door'. It can also be used to state that a certain action is permissible, as in the deontic sentence 'You may (are legally permitted to) park on this street after dark'. In the use of 'must' and 'may' context should usually suffice to distinguish between an imperative and deontic sentence, though there may be borderline cases that may be difficult to resolve.

Let us represent the expressions 'it is obligatory that...' and 'it is permissible that...' by the deontic operators 'O' and 'P'. Then if A! is a sentence schema representing the imperative 'Pay the bill', 'It is obligatory that the bill be paid' or '"Pay the bill" is obligatory' would be represented by OA! and 'It is permissible to pay the bill' would be represented by PA!.[3] It is evident that the terms of the deontic trichotomy can all be defined in terms of the obligation and permission operators. 'A! is indifferent' would be represented by PA!∧P-A!, while 'A! is forbidden' would be represented by -PA! or O-A!. As for the alethic modal operators, the two deontic operators are interdefineable by means of the definitions,

$$\text{i) } OA! =_{df} -P-A!$$
$$\text{ii) } PA! =_{df} -O-A!$$

To obey a prescription is obligatory if and only if it is not permitted to disobey it; while to obey a prescription is permissible if and only if it is not obligatory to disobey it. Parentheses are again used to indicate the scope of the deontic operator. Thus, the parentheses in P(A!∧B!) and O(A⊃B!) indicate that the scopes of the operators include a conjunction and implication. Note that in the latter expression the antecedent of the implication is in the indicative mood. As we saw in the previous section, this is required by the nature of the conditional imperatives the implication is used to represent.

Deontic inferences in which terms recur as constituents can be represented by prefixing the deontic operators before quantificational schemata. Thus, 'Everything ought to be picked up' could be paraphrased by '"Pick up everything" is obligatory'. Its representation is now O∀xP!x, with P for 'pick up...'. Addresses must again be represented where their logical relationship affects the validity of an inference. If we do need to represent addresses, then sentences such as 'Peter ought to pay the bill' and 'Someone ought to pay the bill' will be paraphrased by '"Peter, pay the bill" is obligatory' and '"Someone pay the bill" is obligatory'. Their representation is now OR!ab and O∃x(Px∧R!xb), with a for 'Peter', b for 'the bill', P for '...is a person' and R! for

'...pay...'. Where logical relationships between addresses are not essential to validity, they can be ignored in our representation.

Types of Obligation. The deontic terms 'obligatory', 'permissible', 'indifferent', and 'forbidden' can be interpreted in a variety of senses. We now list the main interpretations for the term 'obligatory'. Since the remaining deontic terms can be defined in terms of this one, these interpretations apply to them also.

1. Logical Obligation. We can regard an action as logically obligatory if it is impossible not to obey the prescription requiring that action. Thus, all prescriptions it is impossible to disobey, e.g. those expressed by imperatives of the forms $A! \vee -A!$ and $-\exists x(P!x \wedge -P!x)$, are obligatory in this sense. Here the justification for the prescription is in terms of the "logical compulsion" for performing a certain action. A logically permissible prescription would be one which is logically possible to perform, that is, any perscription that is not a logical contradiction.

2. Moral Obligation. When we say 'Promises ought to be kept' or 'One ought not to harm one's neighbors' we are using 'ought' in the moral sense of a requirement imposed by virtue of certain conditions of life within a social group. Exactly what these conditions are and whether they can be applied to all societies is a source of considerable dispute among ethical philosophers. Morally permissible actions are those not morally forbidden. Morally indifferent actions are those that are neither obligatory nor forbidden.

3. Legal Obligation. This is an obligation imposed by a law or ordinance of a community, e.g. the obligation to register a car, to drive within certain speed limits, or to pay federal taxes on income. Among legal obligations we should include also those stemming from the authority invested by law in certain people, e.g. the obligation of the motorist to obey the policeman and the private to obey the sergeant.

4. Practical Obligation. We often use the word 'ought' to express an obligation that we impose on ourselves relative to some goal and a belief about the means necessary to achieve it. For example, if we say 'The house ought to be built with good materials' we perhaps intend to say that it should be built that way because of our desire to have a durable house that will require little maintenance afterwards and our belief that good materials are necessary for such a house. The action is obligatory as what is believed necessary in order to realize a desired end. A permissible action would be one not in conflict with this goal.

It is common to say of a person that it is his "duty" to do something instead of saying he "ought" to do it, especially in the areas of morality and law. Thus we say 'It is your (moral) duty to help your neighbor' instead of 'You ought to help your neighbor' and 'It is your (legal) duty to pay taxes' instead of 'You ought to pay taxes'. It has been suggested that a person's duties follow from his station or position (the duties of a husband or citizen) and that these duties provide a special kind of justification for deontic 'ought' sentences.[4] Nevertheless, for the purposes of logical evaluation 'It is your duty to...' and 'You ought to...' can be regarded as synonymous.

The obligation and permission operators of deontic logic can be interpreted in at least the four different ways that have been listed. There are, of course, complex relations of interdependence between the various obligations. For example, some (but not all) morally obligatory actions, e.g. not inflicting harm on one's neighbor, are also legally obligatory. It might also be maintained that a moral obligation is a kind of practical obligation imposed by a goal shared by members of a social group. Nevertheless, the types of obligation do seem distinguishable, and should not be confused with another.

As before for alethic modal logic, we can also distinguish between the categorical and conditional deontic modalities. A <u>conditional obligation</u> is one that is imposed relative to the truth of some proposition. Thus B! is conditionally obligatory if there is some indicative A such that $O(A \supset B!)$. For example, one might be legally obligated to pay taxes if earning over $700 or morally obligated to help others if they are in distress. Both would be conditional obligations. A <u>categorical obligation</u> is one that is not conditional on some circumstance, e.g. to act in a manner that can be universalized (Kant's "categorical imperative").

<u>Special Features of Deontic Logic</u>. The structure of deontic logic is almost the same as alethic modal logic, with the deontic schemata OA! and PA! paralleling the alethic schemata $\Box A$ and $\Diamond A$. There are, however, two major differences between the two branches, both arising from the nature of the imperative schemata within the scope of the deontic operators.

1. Conditional Obligations. The first difference arises from the special nature of the conditional imperatives represented by $A \supset B!$, with the antecedent A in the indicative mood and B! in the imperative. The deontic sentences we have considered have been meta-linguistic sentences in which imperatives are named. It seems this must be an essential feature of all deontic sentences, for it does not seem possible for us to name indicatives in these sentences. It makes no sense, for example, to say '"The door is closed" is obligatory' or 'It is obligatory that the door is closed'. Instead, we say 'It is obligatory that the door <u>be</u> closed' or 'The

door ought to be closed', and these sentences, as we have seen, can be paraphrased by '"Close the door" is obligatory'. An expression such as OA, where the obligation operator has within its scope a sentence schema in the indicative mood, must therefore be excluded from the symbolism we use to represent deontic sentences. Similar considerations hold for permission sentences. It is an action of obeying a prescription that is permissible and not the truth of a proposition. Hence, PA! represents a deontic sentence, but not PA.

It follows as a direct consequence of this restriction on deontic schemata that the conditional obligation $O(A \supset B!)$ does not entail $OA \supset OB!$, although in alethic modal logic $\Box(A \supset B)$ does entail $\Box A \supset \Box B$. The reason for this difference is simply that the expression OA is not a deontic schema capable of representing a sentence of the object language.[5] $O(A \supset B!)$ does entail (and is entailed by) $A \supset OB!$, for it is obligatory that B if A if and only if B is obligatory if A, e.g. 'It is obligatory that if the light turns red you stop' is logically equivalent to 'If the light turns red you ought to stop'. This equivalence can be stated in the form of a definition.

$$\text{iii)} \quad O(A \supset B!) =_{df} A \supset OB!$$

In contrast, $\Box(A \supset B)$ is not logically equivalent to $A \supset \Box B$, for if A entails B it does not follow that if A is true B is necessary.

In similar fashion a conditional permission of the form $P(A \supset B!)$ (e.g. 'It is permissible to open the window if it is hot') is equivalent to a sentence of the form $A \supset PB!$ ('If it is hot you may open the window'). Hence the definition,

$$\text{iv)} \quad P(A \supset B!) =_{df} A \supset PB!$$

The obligation and permission operators are distributed in a similar manner over the material equivalence connective. Thus we have $O(A \equiv B!) \Leftrightarrow A \equiv OB!$ and $P(A \equiv B!) \Leftrightarrow A \equiv PB!$.

2. Inferences between Deontic Sentences and Imperatives. The second contrast between alethic and deontic logic is more fundamental. In the modal calculus of Section 34 we took as a primitive rule the rule of necessity elimination that allows us to infer from a necessity schema to the schema within the scope of the necessity operator. This is clearly a valid rule, for if a proposition is necessarily true, then it follows that it is true. Also included as a primitive rule was the rule of possibility introduction that allows us to infer from a sentence schema to a possibility schema in which this schema occurs as operand. This rule is justified by the obvious fact that if a proposition is true, then it follows that it is possible that it is true.

Neither of the corresponding inferences is valid within deontic logic, however. We cannot infer from an obligation sentence of the form OA! to the imperative A!, for the fact that it is true that action is obligatory is no assurance that the action will be performed, that the imperative A! will be obeyed. If men had invariably done what they ought to have done the world would be a much different place than it is now and the religious would have little chance for confession. Nor can we infer from an imperative A! to a deontic sentence of the form PA!. From the fact that a prescription is obeyed it does not necessarily follow that the action of obeying it is permitted. There may be requirements placed upon a person to which he submits, but for which no justification can be given.

Both a decision procedure and calculus can be constructed for deontic logic. As we shall see in the next section, the differences between this procedure and calculus and that for alethic modal logic are due solely to the two special features of deontic logic just discussed.

39. Decision Procedure for Deontic Logic

The Decision Procedure. With but one exception due to the special nature of conditional obligations and permissions the decision procedure for deontic logic parallels that for alethic modal logic developed in Section 33.[1] To evaluate deontic inferences as valid or invalid or the logical status of molecular deontic sentences we again represent the logical form of the inferences or sentences and expand the representing schemata into their distributive normal forms. Here this normal form is a truth function of permission-constituents or P-constituents in which the permission operator has within its scope each of the constituent sentence schemata or their negations. We then construct a truth table whose columns are these P-constituents. If by means of the table it is found impossible for the premisses of an inference with n constituent sentence schemata to be true and the conclusion false for all possible $2^{2^n}-1$ interpretations, the inference is valid; and if a deontic sentence is true for all possible interpretations, it expresses a logical truth.

This procedure may be illustrated by applying it to the inference,

1) You ought to either stop drinking or not drive home.
   ∴ Either you ought to stop drinking or you ought not drive home.

Adopting the same abbreviated symbolism for negation and conjunction as before, the logical form of 1) is

2) $\dfrac{O(A! \vee \overline{B}!)}{OA! \vee O\overline{B}!}$

The expansion of the premiss of 2) into distributive normal form is

3) $O(A! \vee \overline{B}!)$
   $-P\overline{(A! \vee \overline{B}!)}$ (definition i) of O operator)
   $-P(\overline{A}!\overline{B}!)$ (expansion of sentence schema into disjunctive normal form)

That for the conclusion is

4) $OA! \vee O\overline{B}!$
   $-P\overline{A}! \vee -P\overline{\overline{B}}!$ (definition of O operator)
   $-P(\overline{A}!B! \vee \overline{A}!\overline{B}!) \vee -P(A!B! \vee \overline{A}!B!)$ (expansion of sentence schemata)
   $-[P(\overline{A}!B!) \vee P(\overline{A}!\overline{B}!)] \vee -[P(A!B!) \vee P(\overline{A}!B!)]$ (distribution of operator over disjuncts)

There are altogether three P-constituents in the distributive normal forms that result from expansions 3) and 4). The truth table needed to evaluate 1) thus can be abbreviated to include only eight possible interpretations.

① $P(\overline{A}!\overline{B}!)$	② $P(\overline{A}!B!)$	③ $P(A!B!)$	-①	-[② ∨ ①]	∨ -[③ ∨ ②]
T	T	T	F	F	T	F F	T
T	T	F	F	F	T	F F	T
T	F	T	F	F	T	F F	T
T	F	F	F	F	T	T T	F
F	T	T	T	F	T	F F	T
F	T	F	T	F	T	F F	T
F	F	T	T	T	F	T F	T
F	F	F	T	T	F	T T	F

It is readily seen that 1) is an invalid inference, since its premiss is true and conclusion false for the fifth and sixth interpretations.

By this procedure we can establish the following logical relations for the distribution of the deontic operators:

$O(A! \wedge B!) \Leftrightarrow OA! \wedge OB!$
$OA! \vee OB! \Rightarrow O(A! \vee B!)$
$P(A! B!) \Rightarrow PA! \wedge PB!$

The truth table just constructed establishes that $O(A! \vee B!)$ does not entail $OA! \vee OB!$. We can also establish that $PA! \wedge PB!$ does not entail $P(A! \wedge B!)$. Note the parallel between these relations and those for the alethic modal operators stated at the end of Section 33.

When conditional obligations or permissions occur within the inferences or sentences being evaluated we must modify our expansion procedure. Here we utilize definitions iii) and iv) of the previous section in order to replace the schemata $O(A \supset B!)$ and $P(A \supset B!)$ by $A \supset OB!$ and $A \supset PB!$. Thus, the expansion of $P(A \supset B! \lor C!)$ becomes:

$$P(A \supset B! \lor C!)$$
$$A \supset P(B! \lor C!) \quad \text{(by definition iv))}$$
$$A \supset P(B!C! \lor B!\overline{C}! \lor \overline{B}!C!)$$
$$A \supset P(B!C!) \lor P(B!\overline{C}!) \lor P(\overline{B}!C!)$$

Note that after the second step where we replace $P(A \supset B! \lor C!)$ by $A \supset P(B! \lor C!)$ the expansion proceeds as before. The effect of this second step is to produce a "mixed" schema that is a truth function of both sentence schemata and P-constituents.[2] This preliminary step is required also for material equivalences within the scope of the deontic operators. Thus, we would first replace $O(A \equiv B!)$ and $P(A \equiv B!)$ by the logically equivalent $A \equiv OB!$ and $A \equiv PB!$ in expanding them to normal form.

To illustrate the application of the decision procedure to conditional obligations and permissions consider the inference,

5) It is obligatory that if you are issued a ticket you either pay a fine or file an appeal.
You are issued a ticket.
You may not file an appeal.
∴You are obligated to pay a fine.

whose form is

6) $O(A \supset B! \lor C!)$
    A
    $-PC!$
    ────
    $OB!$

The distributive normal form of the first premiss of 6) is reached by the following steps:

$$O(A \supset B! \lor C!)$$
$$A \supset O(B! \lor C!) \quad \text{(by definition iii))}$$
$$A \supset -P\overline{(B! \lor C!)}$$
$$A \supset -P(\overline{B}!\overline{C}!)$$

The distributive normal form of the third premiss is $-[P(B!C!) \lor P(\overline{B}!C!)]$, while that of the conclusion is $-[P(\overline{B}!C!) \lor P(\overline{B}!\overline{C}!)]$. There are four constituents in the distributive normal form, three of them P-constituents, and the other the indicative sentence schema A. The truth table thus requires four columns.

A	P(B!C!)②	P(B̄!C!)③	P(B̄!C̄!)④	A⊃-④	-(②∨③)	-(③∨④)
T	T	T	T	F	F T	F T
T	T	T	F	T	F T	F T
T	T	F	T	F	F T	F T
T	T	F	F	T	F T	T F
T	F	T	T	F	F T	F T
T	F	T	F	T	F T	F T
T	F	F	T	F	T F	F T
T	F	F	F	T	T F	T F
F	T	T	T	T	F T	F T
F	T	T	F	T	F T	F T
F	T	F	T	T	F T	F T
F	T	F	F	T	F T	T F
F	F	T	T	T	F T	F T
F	F	T	F	T	F T	F T
F	F	F	T	T	T F	F T
F	F	F	F	T	T F	T F

The truth table shows 5) to be valid, since there is no interpretation under which the premisses are all true and the conclusion false.

Exercises. Evaluate the following inferences as valid or invalid.
1. Either you ought to pay the bill or you ought to return the package. Therefore, you ought to either pay the bill or return the package.
2. It is forbidden to either break a promise or tell a lie. Hence, it is forbidden to break a promise and forbidden to tell a lie.
3. Whether or not you pay your taxes now is indifferent. But you must pay them in April. Hence you must either pay your taxes now or by April.
4. You may speed on this highway, but not on the side street. Hence, it is either permissible to speed on the highway or permissible to not speed on the side street.
5. You ought to pay the bill if you bought the merchandise. But you didn't buy the merchandise. Therefore, you should not pay the bill.
6. You may break the promise and fail to come only if you have an excuse. But you don't have an excuse. Hence, you must come.
7. It is forbidden to either search the house or make an arrest unless you have a warrant. You have a warrant. Therefore, you may search the house.
8. You ought not leave Sheilah without giving her your address unless you don't want to see her again. But you are forbidden to give her your address. Hence, if you want to see her again you should not leave Sheilah.

40. The Deontic Calculus

The rules of the deontic calculus include those of the predicate calculus as modified for application to imperatives plus rules governing the deontic operators. The deontic operator rules differ from those governing the alethic modal operators with respect to the two distinctive features of deontic logic noted at the end of the previous section.[1]

Obligation Rules. The rule of obligation introduction (O intro) parallels that of necessity introduction in the modal calculus.

R21. Obligation Introduction: From the proof within an obligation sub proof of an imperative schema ϕ we may infer a deontic schema of the form Oϕ.

```
  | .
O | .
  | .
  |_φ
  ──
   Oφ
```

By an "obligation sub proof" we mean a sub proof into which can be reiterated only those imperative schemata prefixed by an obligation operator. After reiteration the operator must be dropped, as for necessity sub proofs in the modal calculus (cf. Section 34). Reiteration into obligation sub proofs is thus restricted by a Principle of Deontic Reiteration analogous to the Principle of Modal Reiteration.

> Principle of Deontic Reiteration: Only imperative schemata prefixed by the obligation operator can be reiterated into obligation sub proofs, and after reiteration the operator must be dropped.

Notice that there is no restriction against indicative schemata being reiterated into obligation sub proofs.

The rule of obligation elimination must be reformulated in such a way as to not permit us to be able to infer from a deontic sentence to a conclusion that is the imperative named within it.

R22. Obligation Elimination: From a deontic schema of the form Oϕ we may infer the imperative schema ϕ provided ϕ is inferred from Oϕ within an obligation sub proof.

```
   | Oφ
 O |──
   | φ
```

This rule has the effect of enabling us to remove the deontic operator in a way needed to conduct proofs. At the same time, however, the restriction that the operand ϕ be inferred from $O\phi$ only within an obligation sub proof prevents what we saw in Section 38 to be the invalid inference from an obligation sentence to an imperative. As for the general sub proofs of the predicate calculus (cf. Section 26), we suspend interpretation within obligation sub proofs, and hence we cannot infer within them from the truth of a proposition to the obeying of a prescription.

The following proof illustrates the use of the two rules just formulated.

```
1 |OA! ∧ OB!        p
2 |OA!              1, ∧ elim
3 |OB!              1, ∧ elim
4   | A!            2,r
5  O| B!            3,r
6   | A! ∧ B!       4,5, ∧ intro
7 |O(A! ∧ B!)       4-6, O intro
```

The conclusion in step 7 is justified by the rule of obligation introduction from the sub proof 4-6 whose conclusion is the imperative $A! \land B!$.

The special nature of conditional obligations requires an additional rule for distributing the deontic operator over the terms of an imperative implication.

R23. <u>Obligation Distribution</u>: <u>From a conditional obligation of the form $O(\phi \supset \psi)$ we may infer a schema of the form $\phi \supset O\psi$, where ϕ represents an indicative schema.</u>

$$\frac{O(\phi \supset \psi)}{\phi \supset O\psi}$$ where ϕ represents an indicative schema.

Within the calculus we cannot infer $OA \supset OB!$ from $O(A \supset B!)$, since the deontic operators are understood as including within their scope only schemata in the imperative mood.

<u>Permission Rules</u>. A restriction similar to that for R22 must be imposed on the introduction of the permission operator.

R24. <u>Permission Introduction</u>: <u>From an imperative ϕ we may infer a deontic schema of the form $P\phi$, provided $P\phi$ is inferred from ϕ within an obligation sub proof.</u>

$$O \left| \frac{\phi}{P\phi} \right.$$

The restriction that Pϕ only be inferred from ϕ within an obligation sub proof prevents the invalid inference from the performance of a certain action to that action's permissibility (cf. Section 38), while at the same time permitting the introduction of the permission operator in uninterpreted steps of a proof.

R25. <u>Permission Elimination</u>: <u>From a deontic schema Pϕ and an obligation sub proof showing that the imperative schema ϕ leads to a deontic schema ψ we may infer ψ.</u>

$$\begin{array}{l} P\phi \\ O \left| \begin{array}{l} \phi \\ \vdots \\ \psi \end{array} \right. \\ \hline \psi \end{array}$$
where ψ must be a deontic schema.

The restriction that ψ be a deontic schema prevents our being able to infer from PA! to A!. Because an action is permitted it does not necessarily follow that an imperative prescribing it is obeyed.

In addition, we again need a rule for distributing the permission operator over an imperative implication. This time, however, the rule required moves the operator from the consequent to a position outside parentheses.

R26. <u>Permission Distribution</u>: <u>From an implication schema of the form $\phi \supset P\psi$ we may infer a deontic schema of the form P($\phi \supset \psi$), where ϕ is an indicative schema.</u>

$$\frac{\phi \supset P\psi}{P(\phi \supset \psi)}$$
where ϕ represents an indicative schema.

The following proof illustrates the use of the permission rules in conjunction with those for the obligation operator.

```
1  P(B! ∨ C!)          p
2  O(A ⊃ -B!)          p
3  A                   p
4  A ⊃ O-B!            2,O dist
5  O-B!                3,4,mp
6      ⌈B! ∨ C!        p
7   O  │ B!            5,r
8      │ C!            6,7,d.s.
9      ⌊PC!            8,P intro
10 PC!                 1,6-9,P elim
```

The proof is constructed in accordance with a strategy analogous to the Possibility Strategy of the modal calculus. After distributing the obligation operator in accordance with the rule of obligation distribution and inferring O-B! by modus ponens in step 5 we construct an obligation sub proof headed by the schema within the scope of the permission operator in the first premiss and terminating in the desired conclusion. The conclusion is then inferred by the rule of permission elimination.

From the fact that a certain action is obligatory it clearly follows that that action is permissible. Because of the restrictions in the rules of obligation elimination and permission introduction justifying this inference by the calculus requires an additional rule. We shall call it the "obligation-permission" rule.

R29. Obligation-Permission: From a deontic schema of the form Oϕ we may infer a schema of the form Pϕ.

$$\frac{O\phi}{P\phi}$$

Where we wish to infer a conclusion of the form Pψ from a premiss of the form Oϕ, we first infer Pϕ from Oϕ by the obligation-permission rule. We then proceed to prove the conclusion Pψ from Pϕ by the rule of permission elimination. The following proof illustrates this procedure.

```
1    O(A⊃C!)           p
2    P(A⊃C!)           1, O-P
3       A⊃C!           p
4          A∧B         p
5          A           4, ∧ elim
6    O     A⊃C!        3, r
7          C!          5,6 m.p.
8       A∧B⊃C!         4-7, ⊃ intro
9       P(A∧B⊃C!)      8, P intro
10   P(A∧B⊃C!)         2,3-9, P elim
```

Note how we apply the obligation-permission (O-P) rule to infer step 2, and then proceed to apply the rule of permission elimination.

The complication in the rules of the deontic calculus is due to the special features of deontic logic noted in Section 38. The nature of conditional obligations and permissions requires rules R23 and R26 for distributing the operators over implication. The fact that there are no valid immediate inferences between deontic propositions and prescriptions requires the special form for the rules of obligation elimination and permission introduction. It also requires the addition of R29, the rule of obligation-permission. For without this rule, it would be impossible to infer from an obligation to a permission within the calculus.

Application of the Calculus to Inferences. As for other branches of the natural deduction calculus the deontic calculus provides an alternative means of establishing the validity of inferences. With it we can establish, for example, the validity of the inference,

> Either you ought to pay the fine or you ought to return the money.
> But you are forbidden to return the money.
> ∴ You may pay the fine.

The proof of its validity is

```
 1 | OA! ∨ OB!      p
 2 | O-B!           p
 3 |  | OA!         p
 4 |  | A!          3, O elim
 5 |  | PA!         4, P intro
 6 |  | OB!         p
 7 |  | B!          6, O elim
 8 |  | -B!         2,r, O elim
 9 |  | A!          7,8 -elim
10 |  | PA!         9, P intro
11 | PA!            1,3-5,6-10, ∨ elim
```

The calculus can also be applied to inferences whose representation requires predicate schemata, variables, and quantifiers. One of these is the inference,

> Someone ought to pick up and burn all the refuse.
> Some refuse is wastepaper.
> ∴ Someone ought to burn some wastepaper.

Its representation is

$$O \forall x(Qx \supset R!x \wedge S!x)$$
$$\underline{\exists x(Qx \wedge Tx)}$$
$$O \exists x(Tx \wedge S!x)$$

with Q for '...is refuse', R! for 'pick up...', S! for 'burn...', and T for '...is wastepaper'. Notice that the address 'someone' in the first premiss and conclusion is not included in the representation, the assumption being that this remains constant. If the address were included in representing this inference, the first premiss would be represented by $O \exists x[Px \wedge \forall y(Qy \supset R!xy \wedge S!xy)]$, with P for '...is a person' and R! and S! now representing the relational predicates '...pick up...' and '...burn...'. The conditional proof establishing the validity of the above inference requires applying quantifier rules in addition to rules for the deontic operators and connectives.

```
 1 |O ∀x(Qx ⊃ R!x ∧ S!x)        p
 2 | ∃ x(Qx ∧ Tx)                p
 3 |    | ∃ x(Qx ∧ Tx)           2,r
 4 |    |   |Qx ∧ Tx             p
 5 |    |   |∀x(Qx ⊃ R!x ∧ S!x)  1,r
 6 |    |   |Qx ⊃ R!x ∧ S!x      5, ∀ elim
 7 |    | x |Qx                  4, ∧ elim
 8 | O  |   |R!x ∧ S!x           7,6, m.p.
 9 |    |   |Tx                  4, ∧ elim
10 |    |   |S!x                 8, ∧ elim
11 |    |   |Tx ∧ S!x            9,10, ∧ intro
12 |    |   | ∃ x(Tx ∧ S!x)      11, ∃ intro
13 |    | ∃ x(Tx ∧ S!x)          3,4-12, ∃ elim
14 |O ∃ x(Tx ∧ S!x)              3-13, O intro
```

The proof is constructed by setting up the first sub proof in accordance with a strategy analogous to the Necessity Strategy of the modal calculus and then applying the Existential Strategy to the existential premiss after it has been reiterated into this sub proof. Note that the second premiss is reiterated into the obligation sub proof at step 3, though it is not prefixed by the obligation operator. Reiteration is restricted within the deontic calculus only for imperative schemata.

We must take the same precaution in combining quantifiers and deontic operators as we did for modal operators in Section 34. We must not, for example, represent the sentence,

    1) Everything ought to be picked up

by

    2) ∀xOP!x

For this representation would require 1) to be paraphrased by

    3) Everything is such that it is obligatory that it be picked up.

Since we have taken deontic sentences to be sentences of the meta-language in which imperatives are mentioned, 3) is in turn equivalent to

    Everything is such that 'pick it up' is obligatory.

But now the pronoun 'it' within the imperative cannot perform its proper role of referring back to the subject 'thing', since it is isolated by quotation marks. To avoid this 1) should be paraphrased by

It is obligatory that everything be picked up

and represented by

$$O \forall x P!x$$

The general rule to follow here is similar to that for the modal operators: deontic operators may include within their scope quantificational schemata, but must not occur within the scope of quantifiers. Suitable restrictions placed on the quantifier rules similar to those of Section 34 for modal operators would prevent the illegitimate combinations from being inferred.

Exercises.

I. Prove the following by means of the deontic calculus.

1. $O(A! \wedge B!) \vdash OA! \wedge OB!$
2. $O(A \supset B!), A \vdash OB!$
3. $O(A \supset B!), O-B! \vdash -A$
4. $O(A \wedge B \supset C!), A \vdash O(B \supset C!)$
5. $O(A \supset B! \vee C!), O-B! \wedge O-C! \vdash -A$
6. $P(A! \vee B!) \vdash PA! \vee PB!$
7. $PA! \vee PB! \vdash P(A! \vee B!)$
8. $OA! \vdash -P-A!$
9. $PA! \vdash -O-A!$
10. $O(A! \wedge B!) \vdash P(A! \vee B!)$
11. $A \supset OB! \vdash O(A \supset B!)$
12. $P(A \supset B!) \vdash A \supset PB!$
13. $O\forall x P!x \vee O\forall x Q!x \vdash O\forall x (P!x \vee Q!x)$
14. $P\exists x(P!x \wedge Q!x) \vdash P\exists x P!x \wedge P\exists x Q!x$
15. $O\forall x(Px \supset Q!x), Pa \vdash OQ!a$
16. $P\exists x(Px \wedge Q!x) \vdash \exists x Px \wedge P\exists x Q!x$
17. $O\forall x \forall y R!xy \vdash P\exists y R!ay$
18. $PR!ab \vdash P\exists x \exists y R!xy$
19. $O\forall x(Px \supset \forall y R!xy) \vdash \forall x Px \supset P\exists x \exists y R!xy$
20. $P\exists x(Px \wedge \forall y R!xy) \vdash P\exists x[(Px \vee Qx) \wedge \exists y R!xy]$

II. Demonstrate the validity of the following inferences by means of the deontic calculus. Note that this requires representing addresses in inferences 5-8. Where there are implied premises, they are written in parentheses.

1. Inferences 1-4 and 6 of the exercise at the end of Section 39.
2. We ought to return all the furniture. But we may keep this book case. Therefore, this book case is not furniture.
3. You may keep this book case. This book case is a piece of furniture. Therefore, you may keep some piece of furniture.

4. It is obligatory to repay all debts. There is at least one debt. Hence, it is permissible to repay some debt.

5. You ought to accept this gift. (You are a person.) Therefore, someone ought to accept something.

6. Every American ought to respect the flag. Therefore, every American in Kansas ought to respect the flag.

7. Everyone is permitted to choose at least one prize and keep it. (John is a person.) Therefore, John is permitted to choose at least one prize.

8. Everyone ought to respect every flag. (There are persons and there are flags.) Hence, someone is permitted to respect some flag.

NOTES FOR CHAPTER VIII

Section 35.

1. For a brief survey of the historical development of logic as applied to imperatives see Rescher [Commands], pp. 5-7. This work also includes an extensive bibliography of the literature.

2. The sentence-radical corresponds to what Hare terms in [Morals], Ch. 2 the "phrastic" of a sentence, while the mood-determiner is his "neustic". The phrastic Hare regards as a participial phrase, and hence 3) becomes in his account 'Mailing of the letter'. The indicative neustic is 'yes', the imperative neustic 'please'. An indicative or imperative sentence is formed by adding the neustic to the phrastic. The form corresponding to 1) is thus 'Mailing the letter, yes', while to 2) corresponds 'Mailing the letter, please'.

3. The symbolism adopted here follows the standard symbolism of predicate logic. For an alternative way of representing imperatives in which temporal reference is included see Rescher [Commands].

Section 36.

1. Failure to recognize the special nature of conditional imperatives can be found in Hofstadter and McKinsey [Imperatives]. The result is to overdraw the parallel between indicative and imperative logic.

2. For a more detailed justification of the Principle of Mood Constancy see my [Mood Constancy]. This principle replaces two rules proposed by Hare: 1) no indicative conclusion can be inferred from premisses which cannot be inferred from the indicatives among them alone; and 2) to infer an imperative conclusion there must be at least one imperative premiss. Cf. his [Morals], pp. 28ff.

Section 37.

1. This restriction would be accomplished by revising our formation rules of Section 30 in such a way as to conform to the rules of natural languages. To our primitive symbols we would add '!' as the mood-determiner for the imperative mood. The formation rules then become:
    1) A,B,C, ... are indicative atomic schemata.
    2) If ϕ is an indicative atomic schema, then $\phi!$ is an imperative schema.

3) A schema ϕ with a constituent schema that is imperative is an imperative schema.
4) If ϕ and ψ are schemata, then
   a) $(\phi \supset \psi)$ and $(\phi \equiv \psi)$ are schemata, provided that if ψ is an imperative schema, ϕ is an indicative.
   b) $-(\phi)$ is a schema, provided that if ϕ is of the form $\phi \supset \psi$, ψ is not an imperative schema.
   c) $(\phi \wedge \psi)$ and $(\phi \vee \psi)$ are schemata, provided ϕ and ψ are schemata in the same mood (but only if they are not open schemata).
5) $\phi\nu$ is an indicative schema, where ϕ is a predicate schema and ν a sequence of individual schemata or variables (or both).
6) If $\phi\nu$ is an indicative schema, then $\phi!\nu$ is an imperative schema.
7) If ϕ is a schema, then $\forall\mu(\phi)$ and $\exists\mu(\phi)$ are schemata.

Section 38.

1. The difference between imperatives and deontic sentences is first noted by Ross [Imperatives]. As Ross points out, since the negations of 'Close the door' and 'You ought to close the door' are fundamentally different, the sentences themselves must differ. For criticisms of those who confuse imperatives with deontic sentences see Castaneda [Imperatives], Secs. 2-7. Whether this criticism is wholly applicable to Hare is questionable. To his credit Hare notes in [Morals], p. 27 (footnote) that "modal imperative logic [deontic logic] is as distinct from the logic of simple imperatives as in the case of the indicative mood."

2. This account of deontic sentences as meta-linguistic sentences in which imperatives are mentioned is due to Castaneda. Cf. his [Imperatives], Secs. 21 and 22. According to Castaneda, the deontic sentence 'x ought to do A' "says in the object language ... what the second-order statement, 'The imperative "x, do A" is necessarily justified in the absolute context of ends, etc.', says in the meta-language of the language of action." He qualifies this, however, by referring to a deontic sentence not as equivalent to a meta-linguistic sentence but as the "image in the material mode of speech of the meta-linguistic statement."

3. Imperative schemata within the scope of modal operators can be found in Fisher [Deontic Logic] and Castaneda [Acts].

4. See Mish'alani [Duty].

5. $O(A! \supset B!)$ is, of course, not a schema within our symbolic language, since no imperative schema can have an imperative antecedent. Cf. the formation rules stated above in the note to Section 37. Both Fisher [Deontic Logic] and

Castaneda [Acts] allow imperative schemata with imperative antecedents within the scope of deontic operators. To do so they must violate the rules of the natural languages the schemata are to represent.

For reasons for forbidding the inference from $O(A \supset B!)$ to $OA \supset OB!$ drawn from examples of "contrary-to-duty imperatives" see Chisholm [Imperatives]. The inference results in drawing a contradiction from propositions all of which can plausibly be accepted as true.

Section 39.

1. The procedure that follows is due to von Wright [Logical Studies], Essay IV, the <u>locus classicus</u> for this branch of modal logic. For von Wright deontic schemata of the form OA represent sentences of the object language, with A representing the name of an act.

2. Since von Wright in [Logical Studies], Essay IV regards $O(A \supset B)$ as representing an object language sentence and places no restrictions on the type of antecedent that may occur within the scope of the operator, he allows the inference from it to $OA \supset OB$.

Section 40.

1. For alternative calculi of the axiomatic variety see Fisher [Deontic Logic] and Castaneda [Acts]. The main differences between the calculus formulated here and these alternatives stem from 1) our requirement that the antecedents of imperative schemata must be indicatives; 2) our restriction that deontic operators include only imperative schemata within their scope; and 3) our prohibition against the inferences from $OA!$ to $A!$ and from $A!$ to $PA!$. The effect of these requirements is to complicate the rules of the calculus, but the resultant gain is in their more accurately conforming to the language in which deontic inferences are formulated and our logical intuitions. The calculus formulated here follows with some modifications that in Fitch [Obligation].

APPENDIX. SUMMARY OF SYNTACTIC PRINCIPLES AND RULES

Listed here are the syntactic principles and rules for each of the branches of the natural deduction calculus developed in the text. Also listed are a selected number of some of the more frequently used derived rules.

A. Sentence Calculus

The sentence calculus consists of one syntactic principle plus ten primitive rules of inference.

Principle of Reiteration: Any step may be reiterated as a later step in the proof in which it occurs or into a sub proof relative to this proof.

R1. Implication Introduction*

$$\frac{\begin{array}{|l} \phi \\ \vdots \\ \psi \end{array}}{\phi \supset \psi}$$

R2. Modus Ponens*

$$\frac{\phi \quad \phi \supset \psi}{\psi}$$

R3. Negation Introduction

$$\frac{\begin{array}{|l} \phi \\ \vdots \\ \psi \\ -\psi \end{array}}{-\phi}$$

R4. Double Negation Elimination

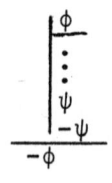

$$\frac{--\phi}{\phi}$$

R5. Conjunction Introduction**

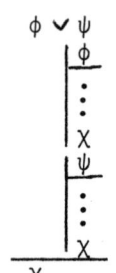

$$\frac{\phi \quad \psi}{\phi \wedge \psi}$$

R6. Conjunction Elimination**

$$\frac{\phi \wedge \psi}{\phi \quad \psi}$$

R7. Disjunction Introduction**

$$\frac{\phi}{\phi \vee \psi \quad \psi \vee \phi}$$

R8. Disjunction Elimination**

$$\frac{\phi \vee \psi \quad \begin{array}{|l}\phi \\ \vdots \\ \chi \end{array} \quad \begin{array}{|l}\psi \\ \vdots \\ \chi \end{array}}{\chi}$$

---

*When applied to imperative schemata, we must assume that ϕ represents an indicative schema and ψ an imperative.

**The conjunction or disjunction to which this rule is applied must consist of constituents in the same mood unless an open schema.

R9. Equivalence Introduction*

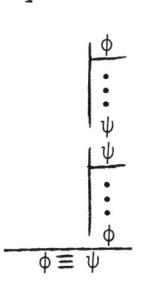

$$\begin{array}{|l}
\phi \\
\vdots \\
\psi \\
\hline
\psi \\
\vdots \\
\phi \\
\end{array}$$
$$\overline{\phi \equiv \psi}$$

R10. Equivalence Elimination**

$$\frac{\phi \equiv \psi}{\phi \supset \psi}$$
$$\psi \supset \phi$$

DR1. Hypothetical Syllogism†

$$\frac{\phi \supset \psi}{\psi \supset \chi}$$
$$\overline{\phi \supset \chi}$$

DR2. Double Negation Introduction

$$\frac{\phi}{--\phi}$$

DR3. Modus Tollens

$$\frac{\phi \supset \psi}{-\psi}$$
$$\overline{-\phi}$$

DR4. Disjunctive Syllogism

$$\frac{\phi \vee \psi}{-\phi}$$
$$\overline{\psi}$$

DR5. Negation Elimination

$$\frac{\phi}{-\phi}$$
$$\overline{\psi}$$

DR6. Contraposition

$$\frac{\phi \supset \psi}{-\psi \supset -\phi}$$

DR7. Negation Disjunction Elimination

$$\frac{-(\phi \vee \psi)}{-\phi \wedge -\psi}$$

DR8. Negation Implication Elimination†

$$\frac{-(\phi \supset \psi)}{\phi \wedge -\psi}$$

---

*When applied to imperative schemata this rule must be reformulated as R9': If $\phi \vdash \psi$ and $-\phi \vdash -\psi$, then $\phi \equiv \psi$, where ϕ is an indicative schema and ψ an imperative schema.

**When applied to imperative schemata this rule becomes R10': $\phi \equiv \psi \vdash \phi \supset \psi$, $-\phi \supset -\psi$, with ϕ an indicative and ψ an imperative schema.

†Cannot be derived when the calculus is applied to imperatives.

B. Predicate Calculus

The predicate calculus consists of the Principle of Reiteration and rules R1-10 of the sentence calculus plus two additional syntactic principles and four primitive rules of inference.

Principle of Restricted Reiteration: No schema in which a free variable occurs may be reiterated into a sub proof general with respect to that variable.

Principle of Substitution: An individual schema must be substituted for a free variable within any proof not a general sub proof for every occurrence of that variable, but substitution for free variables within general sub proofs is not required.

R11. Universal Quantifier Introduction*

$$\frac{\nu \mid \phi\nu}{\forall\mu\phi\mu}$$

R12. Universal Quantifier Elimination*

$$\frac{\forall\mu\phi\mu}{\phi\nu}$$

R13. Existential Quantifier Introduction*

$$\frac{\phi\nu}{\exists\mu\phi\mu}$$

R14. Existential Quantifier Elimination*

$$\frac{\exists\mu\phi\mu \quad \nu\mid\begin{matrix}\phi\nu\\ \vdots\\ \psi\end{matrix}}{\psi}$$

where ψ contains no free occurrences of ν

DR9. Negation Existential Quantifier Introduction

$$\frac{\forall\mu-\phi\mu}{-\exists\mu\phi\mu}$$

DR10. Negation Existential Quantifier Elimination

$$\frac{-\exists\mu\phi\mu}{\forall\mu-\phi\mu}$$

DR11. Mixed Universal Quantifier Distribution

$$\frac{\forall\mu(\psi\vee\phi\mu)}{\psi\vee\forall\mu\phi\mu}$$

DR12. Mixed Existential Quantifier Distribution

$$\frac{\exists\mu(\psi\vee\phi\mu)}{\psi\vee\exists\mu\phi\mu}$$

---

*The expression $\phi\mu$ within the scope of the quantifier may not include a modal operator.

C. Predicate Calculus with Identity

The predicate calculus with identity consists of the rules of the predicate calculus plus two rules governing the identity sign.

R15. Identity Introduction

$$\begin{array}{|l} \;\vdots \\ \;\psi \\ \hline \nu = \nu \end{array}$$

R16. Identity Elimination*

$$\frac{\mu = \nu \quad \phi\mu}{\phi\nu}$$

DR13. Identity Symmetry

$$\frac{\mu = \nu}{\nu = \mu}$$

DR14. Identity Transitivity

$$\frac{\mu = \nu \quad \nu = \rho}{\mu = \rho}$$

D. Alethic Modal Calculus

The alethic modal calculus consists of the rules and principles of the predicate calculus with identity plus an additional reiteration principle and four rules of inference.

Principle of Modal Reiteration: Only schemata prefixed by the necessity operator may be reiterated into necessity sub proofs, and after reiteration the operator must be dropped.**

R17. Necessity Introduction

$$\begin{array}{l} \Box \;\begin{array}{|l} \;\vdots \\ \;\phi \end{array} \\ \hline \phi \end{array}$$

R18. Necessity Elimination

$$\frac{\Box\phi}{\phi}$$

R19. Possibility Introduction

$$\frac{\phi}{\Diamond\psi}$$

R20. Possibility Elimination

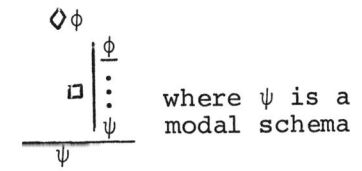

$$\begin{array}{l} \Diamond\phi \\ \Box\;\begin{array}{|l} \phi \\ \vdots \\ \psi \end{array} \\ \hline \psi \end{array}$$ where ψ is a modal schema

---

*When used in conjunction with the modal rules that follow it must be understood that the expression $\phi\mu$ in this rule does not contain a modal operator.

**To form the iterative modal calculus we change this principle to read: "Only schemata prefixed by a modal operator may be reiterated into necessity sub proofs."

E. Deontic Calculus

The deontic calculus consists of the rules and principles of the predicate calculus with identity as modified to apply to imperatives plus the following principle and rules.

Principle of Deontic Reiteration: Only imperative schemata prefixed by the obligation operator may be reiterated into obligation sub proofs, and after reiteration the operator must be dropped.

R21. Obligation Introduction

$$\dfrac{O\ \begin{vmatrix}\ \vdots\\ \phi\end{vmatrix}}{O\phi}$$

R22. Obligation Elimination

$$O\ \begin{vmatrix}O\phi\\ \phi\end{vmatrix}$$

R23. Obligation Distribution

$$\dfrac{O(\phi \supset \psi)}{\phi \supset O\psi} \quad \text{where } \phi \text{ is an indicative schema and } \psi \text{ an imperative.}$$

R24. Permission Introduction

$$O\ \begin{vmatrix}\phi\\ P\phi\end{vmatrix}$$

R25. Permission Elimination

$$\dfrac{P\phi \quad O\ \begin{vmatrix}\phi\\ \vdots\\ \psi\end{vmatrix}}{\psi} \quad \text{where } \psi \text{ is a deontic schema}$$

R26. Permission Distribution

$$\dfrac{\phi \supset P\psi}{P(\phi \supset \psi)} \quad \text{where } \phi \text{ is an indicative schema and } \psi \text{ an imperative}$$

R27. Obligation-Permission

$$\dfrac{O\phi}{P\phi}$$

BIBLIOGRAPHICAL REFERENCES

The following list is restricted to works cited in the notes in the text. The abbreviated titles used in the notes are placed to the left of the full titles.

Ackermann, Wilhelm, [Decision Problem] Solvable Cases of the Decision Problem. Amsterdam: North Holland, 1954.

Anderson, Alan Ross and Belnap, Nuel D., [Entailment] "The Pure Calculus of Entailment," Journal of Symbolic Logic, Vol. 2 (1962), pp. 19-52. Reprinted in part in Iseminger [Logic].

Anscombe, G. E. M., [Intention] Intention. Ithaca: Cornell Univ. Press, 1957.

Arnold, B. H., [Algebra] Logic and Boolean Algebra. Englewood Cliffs: Prentice-Hall, 1962.

Ayer, Alfred Jules, [Language] Language, Truth and Logic, 2nd ed. New York: Dover, 1946.

Basson, A. H. and O'Connor, D. J., [Logic] Introduction to Symbolic Logic, 3rd ed. New York: Free Press, 1960.

Belnap, Nuel D., [Tonk] "Tonk, Plonk and Plink," Analysis, Vol. 22 (1962), pp. 130-134. Reprinted in Strawson [Phil. Logic].

Bennett, Jonathan, [Entailment] "Entailment," Philosophical Review, Vol. 78 (1969), pp. 197-236.

Birkhoff, Garrett and MacLane, Saunders, [Survey] A Survey of Modern Algebra, 3rd ed. New York: Macmillan, 1965.

Bochenski, I. M., [History] A History of Formal Logic. Notre Dame: Univ. of Notre Dame Press, 1961.

Carnap, Rudolf, [Necessity] Meaning and Necessity, 2nd ed. Chicago: Univ. of Chicago Press, 1956.

_____, [Empiricism] "Empiricism, Semantics, and Ontology," Appendix A of [Necessity].

_____, [Postulates] "Meaning Postulates," Appendix B of [Necessity].

_____, [Syntax] The Logical Syntax of Language. Trans. by Smeaton. Patterson: Littlefield and Adams, 1959.

Castaneda, Hector Neri, [Imperatives] "Imperatives, Decisions, and 'Oughts': A Logico-Metaphysical Investigation" in Castaneda and Nakhnikian [Morality].

_____, [Acts] "Acts, the Logic of Obligation, and Deontic Calculi," Philosophical Studies, Vol. 19 (1968), pp. 13-26.

Castaneda, Hector Neri and Nakhnikian, George, eds. [Morality] Morality and the Language of Conduct. Detroit: Wayne State Univ. Press, 1963.

Chisholm, Roderick, [Imperatives] "Contrary-to-Duty Imperatives and Deontic Logic," Analysis, Vol. 24 (1963), pp. 33-36.

Church, Alonzo, [Entscheidungsproblem] "A Note on the Entscheidungsproblem," Journal of Symbolic Logic, Vol. 1 (1936), pp. 40-41, 101-102.

Church, Alonzo, [Logic] *Introduction to Mathematical Logic*. Part I. Princeton: Princeton Univ. Press, 1944.

Clarke, David S., Jr., [Mood Constancy] "The Principle of Mood Constancy," *Analysis*, Vol. 30 (1970), pp. 100-103.

Cohen, L. Jonathan, [Diversity] *The Diversity of Meaning*. London: Methuen, 1962.

Cohen, Morris R. and Nagel, Ernest, [Introduction] *An Introduction to Logic and Scientific Method*. New York: Harcourt and Brace, 1934.

Copi, Irving, [Logic] *Symbolic Logic*, 3rd ed. New York: Macmillan, 1967.

Copi, Irving and Gould, James, eds. [Readings] *Contemporary Readings in Logical Theory*. New York: Macmillan, 1967.

Feigl, Herbert and Sellars, Wilfred, eds. [Analysis] *Readings in Philosophical Analysis*. New York: Appleton-Century-Crofts, 1949.

Fisher, Mark, [Deontic Logic] "A System of Deontic Logic," *Mind*, Vol. 71 (1962), pp. 231-236.

Fitch, Frederick, [Logic] *Symbolic Logic*. New York: Ronald, 1952.

_____, [Obligation] "Natural Deduction Rules for Obligation," *American Philosophical Quarterly*, Vol. 3 (1966), pp. 27-38.

Flew, Anthony, ed. [Analysis] *Essays in Conceptual Analysis*. London: Macmillan, 1956.

Frege, Gottlob, [Thought] "The Thought," trans. by Quintons, *Mind*, Vol. 65 (1965), pp. 289-311. Reprinted in Iseminger [Readings].

Godel, Kurt, [Undecidable Propositions] *On Formally Undecidable Propositions of Principia Mathematica and Related Systems*. Trans. by Meltzer with introduction by Braithwaite. Edinburgh: Oliver and Boyd, 1962.

Hare, Richard M., [Morals] *The Language of Morals*. Oxford: Clarendon, 1952.

Henkin, Leon, [Completeness] "The Completeness of the First-Order Functional Calculus," *Journal of Symbolic Logic*, Vol. 14 (1947), pp. 159-166. Reprinted in Hintikka [Mathematics].

Heyting, Arend, [Intuitionism] *Intuitionism*. Amsterdam: North Holland, 1956.

Hilbert, David and Ackermann, Wilhelm, [Logic] *Principles of Mathematical Logic*. Trans. by Hammond, Leckie and Steinhardt. New York: Chelsea, 1950.

Hintikka, Jaako, [Knowledge] *Knowledge and Belief*. Ithaca: Cornell Univ. Press, 1962.

_____, [Sea Fight] "The Once and Future Sea Fight: Aristotle's Discussion of Future Contingents in De Interpretatione IX," *Philosophical Review*, Vol. 73 (1964), pp. 461-492.

_____, ed. [Mathematics] *The Philosophy of Mathematics*. Oxford: Oxford Univ. Press, 1969.

Hofstadter, Albert and McKinsey, J. C. C., [Imperatives] "On the Logic of Imperatives," <u>Philosophy of Science</u>, Vol. 6 (1939), pp. 446-457.

Hughes, G. E. and Crosswell, M. J., [Modal Logic] <u>An Introduction to Modal Logic</u>. London: Methuen, 1968.

Iseminger, Gary, ed. [Readings] <u>Logic and Philosophy: Selected Readings</u>. New York: Appleton-Century-Crofts, 1968.

―――, [Logic] <u>An Introduction to Deductive Logic</u>. New York: Appleton-Century-Crofts, 1968.

Jeffrey, Richard C., [Logic] <u>Formal Logic: Its Scope and Limits</u>. New York: McGraw Hill, 1967.

Kneale, William and Kneale, Mary, [Development] <u>The Development of Logic</u>. Oxford: Oxford Univ. Press, 1962.

Kyburg, Henry E., [Probability] <u>Probability and Inductive Logic</u>. New York: Macmillan, 1970.

Leblanc, Hugues, [Deductive Inference] <u>Techniques of Deductive Inference</u>. Englewood Cliffs: Prentice-Hall, 1966.

Lemmon, E. J., [Sentences] "Sentences, Statements and Propositions," in Williams and Montefiore [Analytical Phil.].

Lewis, Clarence Irving and Langford, Cooper Harold, [Logic] <u>Symbolic Logic</u>, 2nd ed. New York: Dover, 1959.

Linsky, Leonard, ed. [Semantics] <u>Semantics and the Philosophy of Language</u>. Urbana: Univ. of Illinois Press, 1952.

Lukasiewicz, Jan, [Syllogistic] <u>Aristotle's Syllogistic</u>, 2nd ed. Oxford: Oxford Univ. Press, 1957.

―――, [Logic] <u>Elements of Mathematical Logic</u>. Trans. by Wajtasiewicz, 2nd ed. New York: Macmillan, 1963.

Marcus, Ruth Barcan, [Modalities] "Modalities and Intensional Languages," <u>Synthese</u>, Vol. 27 (1962), pp. 303-322. Reprinted in Copi and Gould [Readings].

Mish'alani, J. K., [Duty] "'Duty', 'Obligation', and 'Ought'," <u>Analysis</u>, Vol. 30 (1969), pp. 33-40.

Morris, Charles, [Foundations] "Foundations of the Theory of Signs," <u>Encyclopedia of Unified Science</u>, Vol. 1, No. 2. Chicago: Univ. of Chicago Press, 1938.

Nagel, Ernest, [Ontology] "Logic Without Ontology" in Feigl and Sellars [Analysis].

Nagel, Ernest, and Newman, James R., [Godel's Proof] <u>Godel's Proof</u>. New York: New York Univ. Press, 1958.

Neidorf, Robert, [Deductive Forms] <u>Deductive Forms</u>. New York: Harper and Row, 1967.

Pap, Arthur, [Semantics] <u>Semantics and Necessary Truth</u>. New Haven: Yale Univ. Press, 1958.

Pitcher, George, ed. [Truth] <u>Truth</u>. Englewood Cliffs: Prentice-Hall, 1964.

Prior, Arthur N., [Time] <u>Time and Modality</u>. Oxford: Oxford Univ. Press, 1957.

―――, [Inference-Ticket] "The Runabout Inference-Ticket," <u>Analysis</u>, Vol. 21 (1960), pp. 38-39. Reprinted in Strawson [Phil. Logic].

―――, [Logic] <u>Formal Logic</u>, 2nd ed. Oxford: Oxford Univ. Press, 1962.

Quine, Willard Van Orman, [Notes] "Notes on Existence and Necessity," *Journal of Philosophy*, Vol. 40 (1943), pp. 113-127. Reprinted in Linsky [Semantics].
———, [Logic] *Mathematical Logic*, rev. ed. Cambridge, Mass.: Harvard Univ. Press, 1951.
———, [Methods] *Methods of Logic*. New York: Holt, Rinehart, and Winston, 1959.
———, [Word] *Word and Object*. Cambridge, Mass.: M.I.T. Press, 1960.
———, [Logical View] *From a Logical Point of View*, rev. ed. Cambridge, Mass.: Harvard Univ. Press, 1961.
———, [Paradox] *The Ways of Paradox*. New York: Random House, 1966.
Ramsey, Frank P., [Foundations] *The Foundations of Mathematics*. London: Routledge and Kegan Paul, 1931.
Reichenbach, Hans, [Logic] *Elements of Symbolic Logic*. New York: Macmillan, 1947.
Rescher, Nicholas, [Commands] *The Logic of Commands*. London: Routledge and Kegan Paul, 1966.
———, [Many-Valued Logic] *Many-Valued Logic*. New York: McGraw-Hill, 1969.
Resnik, Michael D., [Logic] *Elementary Logic*. New York: McGraw-Hill, 1970.
Rorty, Richard, ed. [Linguistic Turn] *The Linguistic Turn*. Chicago: Univ. of Chicago Press, 1967.
Ross, Alf, [Imperatives] "Imperatives and Logic," *Philosophy of Science*, Vol. 11 (1944), pp. 30-46.
Russell, Bertrand [Denoting] "On Denoting," *Mind*, Vol. 14 (1905), pp. 479-493. Reprinted in Feigl and Sellars [Analysis].
Ryle, Gilbert, [Misleading Expressions] "Systematically Misleading Expressions," *Proceedings of the Aristotelian Society*, Vol. 32 (1931-32), pp. 139-170. Reprinted in Rorty [Linguistic Turn].
———, [Mind] *The Concept of Mind*. New York: Barnes and Noble, 1949.
Strawson, Peter F., [Referring] "On Referring," *Mind*, Vol. 59 (1950), pp. 320-344. Reprinted in Flew [Analysis].
———, [Truth] "Truth" *Proceedings of the Aristotelian Society*, Supp. Vol. 24 (1950). Reprinted in Pitcher [Truth].
———, [Logical Theory] *Introduction to Logical Theory*. London: Methuen, 1952.
———, [Individuals] *Individuals*. London: Methuen, 1959.
———, ed. [Phil. Logic] *Philosophical Logic*. Oxford: Oxford Univ. Press, 1967.
Toulmin, Stephen F., [Argument] *The Uses of Argument*. Cambridge Univ. Press, 1958.
Whitehead, Alftred N. and Russell, Bertrand, [P.M.] *Principia Mathematica*, 3 vols. Cambridge: Cambridge Univ. Press, Vol. I, 1910; Vol. II, 1912; Vol. III, 1913.

Williams, Bernard and Montefiore, Alan, [Analytical Phil.] *British Analytical Philosophy*. New York: Humanities Press, 1966.

Wittgenstein, Ludwig, [Tractatus] *Tractatus Logico-Philosophicus*. Trans. by Pears and McGuinness. London: Routledge and Kegan Paul, 1961.

Wright, Georg Henrik von, [Modal Logic] *An Essay in Modal Logic*. Amsterdam: North Holland, 1951.

_____, [Logical Studies] *Logical Studies*. New York: Humanities Press, 1957.

_____, [Practical Inference] "Practical Inference," *Philosophical Review*, Vol. 72 (1963), pp. 159-179.

INDEX

A. See Square of opposition
Address, 195-96
Alethic modal calculus, 178-86
Alethic modal logic: defined, 162; decision procedure for, 171-77
Alethic modal operator, 161
All, 71, 75, 78-79, 82. See also Universal quantifier
Although, 18. See also Conjunction
Analyticity: formal, 167; material, 167
Analytic sentence, 166-69
And, 18, 82. See also Conjunction
Antecedent: defined, 21; of imperative conditional, 198. See also Conditional
Any, 82. See also Universal quantifier
Aquinas, Thomas, 29
Argument force, 4
Argument from design, 29-30
Aristotle, 97, 123, 129n
Associative law, 37
Asymmetric relation, 116, 118
Atomic sentence, 19, 73
Atomic sentence schema, 19, 73
Attribute, 69
Attributive term, 70
Axiom, 157

Barbara, 101
Because, 18
Biconditional, 24-25
Binding convention, 25
Bivalence, principle of, 122-27
Boolean algebra, 37
Braces, 25. See also Parentheses
Brackets, 25. See also Parentheses
But, 18. See also Conjunction

Calculus: natural deduction, 45; axiomatic, 157
Carnap, Rudolf, 187n
Castaneda, H. N., 230n
Category term: defined, 71; as introduced subject, 78-80, 85-86, 150; as subject of imperatives, 195
Church, Alonzo, 121
Class: in Venn diagram, 100-102; correlation to predicate, 107
Commutative law, 37
Compatible propositions, 165-66
Completeness, 154, 157-58
Conclusion: of inference, 1; of inductive inference, 11n; of proof, 47
Condition, necessary and sufficient, 25
Conditional: defined, 21; logical, 21-22; counterfactual, 22; as form of inference, 34-35; imperative, 198-99, 215-16. See also Entailment; Material implication
Conjunct, 17. See also Conjunction
Conjunction: defined, 17-18; sign, 17-18; rules for, 55-56; equivalence to universal sentence, 91-92; in 3-valued logic, 125; alternative definitions of, 154-56; of imperatives, 193-205
Conjunction elimination, rule of, 56, 205
Conjunction introduction, rule of, 55, 205
Conjunctive normal form, 43n
Connective, logical: defined, 18; main, 20-21
Consequent: defined, 21; of imperative, 198. See also Conditional
Consistency: of rules and axioms, 158; of propositions, 165-66. See also Compatible propositions
Contingency: of events, 124
Contingent proposition, 34, 113, 161-62, 166
Contradiction: defined, 33-34; in square of opposition, 76-77, 97; in predicate logic, 113
Contraposition, 36, 60
Contrary propositions, 77, 97
Conventionalism, 124
Converse relations, 117, 143-44,

169-70
Conversion, 97-98
Copula, 97

Decision procedure: for inferences, 27-31; general, 33-36; effective, 62; for finite domains, 91-96, 118-19; for infinite domains, 94-96, 119-21; for syllogism, 99-103; for monadic predicates, 110-15; for relational predicates, 116-21; for alethic modal logic, 171-77; for imperatives, 192-96; for deontic logic, 217-20
De dicto modality, 163, 183-86. See also Meta-language
Deductive logic, 2-4
Definite description, 69-70, 150
Definition, 26, 154-55
DeMorgan's laws, 36, 37
Deontic calculus, 221-27
Deontic logic, 164, 211
Deontic operator, 213
Deontic sentence, 211-15
De re modality, 162-63, 183-86. See also Object language
Descriptions, theory of, 150
Discourse, 1
Disguise: of inference form, 30-31; of sentence structure, 81-82; of analytic sentence, 168-69
Disjunct, 18. See also Disjunction
Disjunction: defined, 18-19; inclusive, 18; exclusive, 18; rules for, 56-57; strategy, 57; equivalence to existential sentence, 91-92; in 3-valued logic, 125; of imperatives, 193, 205
Disjunction elimination, rule of, 56
Disjunction introduction, rule of, 56, 61, 205
Disjunctive normal form: of sentence schema, 37-40; of open schema, 107
Disjunctive syllogism, 59
Distributive law, 37
Distributive normal form: of predicate schema, 104-9; of modal schema, 173-77; of deontic schema, 217-20
Domain: defined, 75; non-emptiness, 77-78, 94, 105, 111; finite, 91-94, 118-19, 195; infinite, 94-96, 119-21, 195
Double negation: law of, 36; rules for, 53, 59
Double negation elimination, rule of, 53-54
Double negation introduction, rule of, 59
Duty, 215. See also Obligation
Dyadic predicate. See Relational predicate

E. See Square of opposition
∃-constituent. See Existence-constituent
Entailment: defined, 35, 165; its transitivity, 61; paradoxes of, 176
Enthymeme, 30-31, 119, 121, 169, 170
Equivalence. See Logical equivalence; Material equivalence
Equivalence relation, 117
Eternal sentence, 11
Evaluation. See Decision procedure
Event, 162
Every. See All
Excluded middle, law of, 36, 122, 126
Existence-constituent, 104-6, 126-27
Existential commitment, 98
Existential quantifier: representing function, 76-77, 80; rules for, 135-36; strategy, 137; applied to imperatives, 194-95, 200-201
Existential quantifier elimination, rule of, 135
Existential quantifier introduction, rule of, 135
Existential sentence, 91-92
Extension: defined, 69; relation to Venn diagram, 100-101
Extensional context, 189n

Falsity. See Truth

Fermat's Last Theorem, 163
Figure, 99-100
Fitch, Frederick, 67n, 188n
Forbidden action: in deontic trichotomy, 164; representation of, 213
Formal implication, 42n. See also Entailment
Formation rules: for predicate calculus, 156-57; for imperative calculus, 229n, 230n
Frege, 12n

General decision procedure. See Decision procedure
Godel, Kurt, 158, 160n

Hare, R. M., 229n
Hypothetical force, 5
Hypothetical syllogism, 45

I. See Square of opposition
Idempotent law, 37
Identity: in Aristotelian logic, 100-101; sign, 147; defined, 147-52; of indiscernibles, 148; as relation, 148-49; rules for, 151-52
Identity, law of, 36
Identity elimination, rule of, 151
Identity introduction, rule of, 151
Imperative sentence: structure of, 190-92, 194-96, 197-202; distinguished from deontic sentence, 211-13
Implication. See Entailment; Material implication
Implies. See If
Impossibility: of event, 124; of proposition, 162, 166. See also Contradiction
Incompatibility: as interpreting stroke function, 26; as modal relation, 166. See also Contrary propositions
Independence: of rules and axioms, 158
Indeterminacy, 123
Indeterminate value, 123, 124-27
Indicator word, 7, 10
Indifferent action: in deontic trichotomy, 164; representation of, 213
Inference: defined, 1; practical, 2; inductive, 2; weak, 2; strong, 2-3; defeat of, 3; form, 14; immediate, 97-99; imperative, 190; mixed, 197-203
Intension, 69
Intensional context, 189n
Interpretations, possible: use in defining connectives, 16, 17, 19, 23-24; of molecular sentence, 20, 34; of atomic sentence constituents, 28; relation to disjunctive normal form, 40; for finite domain, 93; of ∃-constituents, 104-6, 120-21; in 3-valued logic, 126; of possibility constituents, 172-73; of P-constituents, 217. See also Decision procedure
Intransitive relation, 117
Invalidity, 4. See also Validity
Iteration: of modal operator, 182-83

Juxtaposition, 37, 106, 173

Law, logical, 36
Leibniz, 148
Lewis, C. I., 42n, 187n, 188n
Liar's paradox, 10, 12n
Logic: as theory of inference, vii, 3-4; as normative science, 1
Logical constant, 17
Logical equivalence: defined, 36; between open schemata, 108
Logical form, 14, 28-31, 46. See also Disguise
Logical word, 15
Lukasiewicz, 123, 124, 125, 129n

Major term, 99
Material equivalence: defined, 24-25, 26; sign, 24-25; rules for, 64-65, 207; strategy, 65; imperative, 199, 207; in deontic logic, 216
Material equivalence elimination, rule of, 64, 207
Material equivalence introduction, rule of, 64, 207

Material implication: defined, 21-24; sign, 22-24; paradoxes of, 24; rules for, 46-49; strategy, 51-52; in 3-valued logic, 125-26; imperative, 198-99; in deontic logic, 215-16
Material implication elimination, rule of. See Modus ponens
Material implication introduction, rule of, 49, 61, 205
Mathematical logic, vii, 157-58
Mathematics, 9
Meaning: of sentence, 6-7; of term, 68; specification of, 168-69
Meinong, 150
Mention. See Meta-language
Meta-language: contrast to object language, 8-10; use in stating entailments, 35; and in stating necessity, 164, 183-85; in deontic logic, 212-13, 226-27, 230n
Meta-schema, 48, 131
Middle term, 99
Minor term, 99
Modalities: alethic, 161-62, 165-71, 172; physical, 162-63; epistemic, 163-64; deontic, 164, 213
Modal logic. See Alethic modal logic; Deontic logic
Modal operator. See Alethic modal operator; Deontic operator
Modus ponens, rule of, 46-48, 205
Modus tollens, rule of, 59
Molecular sentence, 19, 73
Molecular sentence schema, 19, 73
Mood constancy, principle of, 202-3
Mood-determiner, 191-93
Morris, Charles, 159n

Necessity: of event, 124; categorical, 166, 167; conditional, 166, 167-69
Necessity elimination, rule of, 180
Necessity introduction, rule of, 178
Necessity operator: rules for, 178-80; strategy, 180

Negation: defined, 15-17; sign, 16-17; rules for, 53-54, 59, 60; strategy, 55; in 3-valued logic, 124-25; internal, 138, 194-95, 201; of imperative, 193, 201. See also Double negation
Negation disjunction elimination, rule of, 60
Negation elimination, rule of, 60
Negation implication elimination, rule of, 60
Negation introduction, rule of, 53
Neustic, 229n
Non-contradiction, law of, 36
Not, 16-17. See also Negation

O. See Square of opposition
Object language: defined, 8-10; applied to physical modalities, 162-63. See also Meta-language
Obligation: of action, 164; representation of, 213; types of, 214-15; categorical, 215; conditional, 215-16, 219-20
Obligation distribution, rule of, 222
Obligation elimination, rule of, 221
Obligation introduction, rule of, 221
Obligation operator: representing function of, 213; rules for, 221-22, 224
Obligation-Permission: rule, 224; strategy, 224
Only if, 25
Open sentence, 73, 81
Operand, 161
Or, 18. See also Disjunction

Paraphrase: of inference, 29-31; of general sentence, 78-83, 85-88
Parentheses: for grouping, 19, 24; binding convention for, 25-26; scope of, 80-81, 87-88, 213; deletion convention for, 85
P-constituent. See Permission constituent
Peirce, Charles, 1
Performance value, 193

Permission, conditional, 216, 219-20
Permission-constituent, 217
Permission distribution, rule of, 223
Permission elimination, rule of, 223
Permission introduction, rule of, 222
Permission operator: representing function of, 213; rules for, 222-25
Phrastic, 229n
Physical modal logic, 162-63
Polish notation, 42n
Possibility constituent, 172-73
Possibility elimination, rule of, 180
Possibility introduction, rule of, 180
Possibility operator: representing function of, 161; rules for, 180-81; strategy, 181
Pragmatics, 154
Predicate: function of, 71-73; monadic, 72, 91, 110, 116; relational, 72, 84-88, 116-21; in Aristotelian logic, 97; of imperative, 191
Predicate calculus: defined, 130; completeness of, 158, 160n; applied to imperatives, 207-9
Predicate logic: defined, 69; second-order, 147, 158
Premiss: of inference, 1; of formal proof, 47; of sub proof, 47
Prescription, 190
Primitive symbol, 156
Proof: defined, 47; conditional, 58, 181-82; categorical, 63-64, 140. See also Sub proof
Proper noun, 69
Proposition: definition of, 6; distinguished from meaning, 7, 11n; as truth bearer, 7-8; as recurring constituent, 13-14

Quantifier: defined, 74; scope of, 80-81, 87-88; combined with modal operator, 183-86; combined with deontic operator, 226-27. See also Existential quantifier; Universal quantifier

Quine, W. V. O., 11n, 42n, 188n, 189n

Recurrence: of sentences, 13-14; of terms, 68-69, 70, 88; of predicates and sentences, 114-15, 138-39; of modal sentences, 176-77
Reduction: of connectives, 26; of syllogisms, 102-3; of mathematics to logic, 158
Referent occasion, 7
Reflexive relation, 117, 148
Reiteration, principle of: in sentence calculus, 50-51; restricted, 130, 137-38; modal, 178-80, 182-83; deontic, 221
Relation: expressed by predicate, 69, 72; modal, 165-66
Relational predicate: of sentence, 116-19; in predicate calculus, 141-44
Replacement, rule of: for schemata, 37-38; for open schemata, 108
Rule: of inference, 44-46; primitive, 45; derived, 58-62
Russell, Bertrand, 150, 159n

Schema: sentence, 14-15, 19; predicate, 73; individual, 73; open sentence, 74, 81; closed sentence, 81; quantificational, 94; modal, 161, 172
Self-reference: paradox of, 10. See also Liar's paradox
Semantics, 154
Sentence: defined, 5-6; singular, 71; general, 71, 194-95
Sentence calculus: defined, 45; applied to imperatives, 204-7
Sentence logic, 13
Sentence-radical, 190-93
Sentence-token, 6
Some, 71-72, 76, 80, 82. See also Existential quantifier
Sortal term, 70
Soundness, 4-5
Square of opposition: for category term subject, 76-77; for

Aristotelian logic, 97; for imperatives, 194-95
Statement, 11n
Strawson, P. F., 11n, 128n, 151, 159
Stroke connective. See Stroke function
Stroke function, 26
Sub-alternate, 77, 97, 98-99
Sub-contrary propositions, 97
Subject: of sentence, 71-73; of inference, 82-83; in Aristotelian logic, 97; of imperative, 190-91
Subordinate proof. See Sub proof
Subordination: between terms, 70-71
Sub proof: defined, 49; general, 130; necessity, 178-79; obligation, 221
Substitution: for variable, 74; principle of, 133-34, 135; in modal contexts, 185-86
Substitution instance, 74
Symmetric relation, 116, 152
Syllogism: form of, 99-100; decision procedure for, 100-103
Syntax, 154
Synthetic a priori, 167, 170-71
Synthetic sentence, 167

Table: truth, 16; performance, 193
Tautology: in sentence logic, 33-36; restricted, 94; unrestricted, 94-95; in predicate logic, 113; in 3-valued logic, 126; in modal logic, 172, 175-76
Term: general, 69-70; singular, 69-70, 150-51
Theorem: in axiomatic calculus, 157; deduction, 159n
Thing: as category term, 71; as subject, 75-76, 78-80, 85-86, 150; as subject of imperative, 200-201. See also Category term
Three-valued logic, 124-27
Token: of sentence, 6; of term, 68
Totally reflexive relation, 117
Transitive relation, 61, 117, 143, 152
Truth: ascribed to propositions, 7-8; function, 16; table, 16; values, 16; predication of, 35, 40

Undecidability: for relational predicates, 119-21; of future, 123, 164
Universal quantifier: representing function, 74-76, 77, 78-80; rules for, 131-32; strategy, 133; applied to imperatives, 200-201
Universal quantifier elimination, rule of, 132
Universal quantifier introduction, rule of, 131
Universal sentence, 91-92
Universe of discourse. See Domain
Unless, 25
Use. See Object language

Validity: definition of, 4; as function of form, 14-15; of rules of inference, 60-62; restricted, 94-95; unrestricted, 94-96; universal, 129n; formal and material, 169-70; of imperative inference, 191-93; radical, 192-93, 202-3; of mixed inference, 197. See also Decision procedure
Variable: individual, 74; as subject, 75-76, 79-80; bound, 80-81; free, 80-81; predicate, 146
Venn diagram, 100-103
von Wright, G. H. See Wright, G. H. von

Wittgenstein, 1
Word, 6
Wright, G. H. von, 128n, 129n, 188n, 231n